**American Energy Choices
Before the Year 2000**

# American Energy Choices Before the Year 2000

Edited by

**Elihu Bergman**
Americans for Energy
Independence

**Hans A. Bethe**
Cornell University

**Robert E. Marshak**
City College of New York

**Lexington Books**
D.C. Heath and Company
Lexington, Massachusetts
Toronto

**Library of Congress Cataloging in Publication Data**

Conference on American Energy Choices Before the Year 2000, City
University of New York, 1978. American energy choices before the
year 2000.

1. Power resources—United States. 2. Energy policy—United States—
Congresses. I. Bergman, Elihu. II. Bethe, Hans Albrecht, 1906-
III. Marshak, Robert Eugene, 1916-      IV. Title.
TJ163.25.U6C63      1978      333.7      78-7122
ISBN 0-669-02398-1                          19Jul'79

Published simultaneously in Canada.

Printed in the United States of America.

International Standard Book Number: 0-669-02398-1

Library of Congress Catalog Card Number: 78-7122

# Contents

Preface     vii

Part I     *Introduction*     1

Chapter 1     **Primary and Alternate Sources of Energy**
*Hans A. Bethe*     3

Part II     *Realities of Energy Conservation*     13

Chapter 2     **Public Policy Issues Relating to Industrial
Cogeneration** *Robert Williams*     15

Chapter 3     **Major Factors of Conservation in Industry**
*Macauley Whiting*     25

Chapter 4     **Citizen Participation in Conservation**
*Shirley Sutton*     31

Chapter 5     **Impact of Conservation on Low- and Fixed-
Income Americans** *Clarke Watson*     35

Chapter 6     **Economic Constraints on Federal Conservation
Targets** *Arnold Moore*     39

Chapter 7     **Refuse Power and Its Immediate Impact on
Energy Conservation** *K.S. Feindler*     45

Part III     *Coal as a Source of Energy*     47

Chapter 8     **Expanding the Use of Coal** *Arthur M. Squires*     49

Chapter 9     **Coal Production and Protection of the
Environment** *Carl Bagge*     71

Chapter 10     **Capitalization and Financing of Coal-Fired
Generating Plants** *Richard Disbrow*     77

Chapter 11     **Development of Federal Coal Resources**
*Frederick N. Ferguson*     81

*Part IV*          *Uranium as a Source of Energy*                    87

✳Chapter 12        **Electric Power Needs of the 1980s**
                   *Aubrey Wagner*                                     89

Chapter 13         **Alternative Nuclear Fuel Cycles**  *Charles Till*      95

Chapter 14         **The Case for the Plutonium Breeder**
                   *John Simpson*                                      109

Chapter 15         **Disposal of High-Level Nuclear Wastes**
                   *Fred Donath*                                       115

Chapter 16         **The Fusion Hybrid**  *Hans A. Bethe*              131

*Part V*           *Conclusion*                                       141

Chapter 17         **Bankrupt Energy Policy: The Abdication of**       143
                   **American Leadership**  *Robert R. Nathan*

                   **About the Contributors**                          151

                   **About the Editors**                              152

# Preface

The energy conferences that have proliferated since 1973 have dealt with some combination of the following three agendas: (1) scientific and technological descriptions of possibilities for adding to the world's energy supply from known and not so well-known sources; (2) an economic analysis of energy requirements projected on a global or individual country scale; and (3) statements of ideal or preferred energy policy alternatives. While adding considerably to our understanding of energy issues, these conferences have not provided an adequate framework in which to consider energy policy issues in pragmatic, as contrasted to descriptive and prescriptive, terms. With the focus of the energy policy discourse shifting from the scientific to the political arena, there is a need for more systematic treatment of the "now and how" of energy policy alternatives. It is this requirement that the conference proceedings reported in this book propose to address.

City College of New York (CCNY), in collaboration with Americans for Energy Independence (AFEI), organized the policy conference which produced these proceedings, to deal with the realities of medium-term energy alternatives in the United States. By interfacing political with technological knowledge, it was envisioned that the conference would yield some realistic assessment of how the national energy requirement might be met before the year 2000.

The unique contribution of the CCNY/AFEI conference was thus its preoccupation with energy policy alternatives in the United States in a short/intermediate time frame, during which the less known and untested sources of energy will not be available for significant exploitation. The conference was designed to make the energy policy discourse more realistic by exploring prospects for meeting national energy requirements with known resources and technologies. To do so, the conference agenda focused on three principal resources capable of making increasingly significant contributions to national energy needs: conservation, coal, and uranium.

During recent years, City College of New York, in part to fulfill its mission as an urban public university, has sponsored an annual conference on a critical public policy issue. The 1977 conference considered the question of a national policy for urban America. For 1978, CCNY sought the collaboration of Americans for Energy Independence, a nonprofit public interest coalition based in Washington, as cosponsor for a discussion of national energy policy.

The conference on "American Energy Choices before the Year 2000" took place at the City University of New York Graduate Center in New York City on January 13-14, 1978, under the joint chairmanship of Dr. Robert Marshak, President of CCNY, and Professor Hans Bethe, chairman of the board of AFEI.

We are indebted to the conference participants and the staff of CCNY and AFEI for their contributions to a worthwhile undertaking; and to the Ford Foundation, the Morton Globus Fund of CCNY, and the Sloan Foundation for their support of the conference and publication of these proceedings.

*Elihu Bergman*
*Hans A. Bethe*
*Robert E. Marshak*

# Part I
# Introduction

# 1

# Primary and Alternate Sources of Energy

*Hans A. Bethe*

Any discussion of the energy problem should start with an assessment of the sources of the type of energy which we are using most.

Figure 1-1 is due to data by Dr. King Hubbert, formerly of the U.S. Geological Survey. For many years Dr. Hubbert has predicted the available resources of oil in this country and in the world. He's a good prophet. For the United States he predicted that oil production would peak in 1970. In fact, it peaked in 1972.

He predicted the amount of oil resources which we might still have available at this point. The geological survey which first opposed him vigorously has finally come down to exactly his prediction. Figure 1-1 is Hubbert's prediction for the world production of oil. Up to 1975 there are actual data. The curve with a high maximum is an extrapolation of this, assuming that consumption will continue to expand as before. If this curve turns out to be true, then just before the year 2000 we shall face a catastrophe: World oil production will reach a peak and therefore will decline inexorably.

If we have great good luck, then the people of the world will save oil. And if such is the case, we may go along the lower curve, which is purely my hopeful imagination; but at least for the last few years we have gone along this lower curve. In this case, the oil of the world will last about 70 years rather than 50.

But, one should keep in mind that any new energy source, and even the installation of plants using old energy sources, takes a very long time. So, indeed, if production reaches a peak in 1997, as the upper curve predicts, then it is high time that we do something about it.

Quite similar to figure 1-1 is the recorded import of treasure—gold and silver—from the New World into Spain. This went along exactly the same bell-shaped curve as oil is likely to do. It reached a maximum, and after that it declined. The most interesting thing, however, is that Spain went bankrupt not at the peak of the curve but at the inflection point, that is, at the point where the slope began to decrease. We are at exactly that point today with regard to oil, and we must watch out that we don't go bankrupt at this point.

Spain went bankrupt because it expected too much in further treasure, just as the American people and the people of the world expect to be able to produce more oil than is realistic. There is something called tertiary oil recovery, which means stimulating old wells in certain ways. This procedure may stretch out oil resources an extra ten years at higher prices. One cannot ask for availability of oil and low prices at the same time.

3

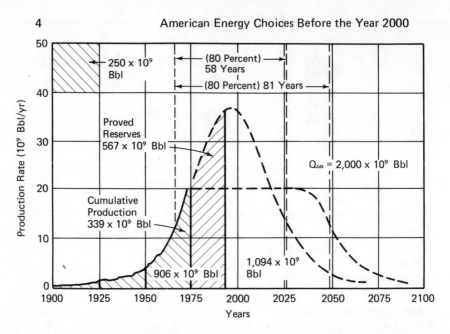

Source: Hubbert, M. King, "World Oil and Natural Gas Reserves and Resources," Figure XIX-3, page 644, in *Project Interdependence: U.S. and World Energy Outlook through 1990*, Congressional Research Service, Library of Congress. (U.S. Government Printing Office, Washington, 1977).

**Figure 1-1.** Estimate of Complete Cycle of World Oil Production.

Now, for the twentieth century, there are only three major choices open to the United States, the industrial world, and some of the developing countries: (1) conservation of oil and natural gas, (2) coal, and (3) nuclear energy. For this reason, the conference has been framed in the way Dean Lustig has mentioned, namely that we discuss these three topics.

Now, to partially justify this, let's discuss some of the alternative sources of energy. Most prominent in the mind of the people is solar energy, and two kinds of solar energy have to be distinguished, namely (1) heating and (2) the use of solar energy for large-scale energy production, like electricity production. These two applications are completely different. Solar heating is almost economically feasible, while the other application is not, for the foreseeable future.

Table 1-1 shows the cost of solar heating installation in dollars per square foot. The information has been provided by Dr. Balcomb, who is in charge of solar energy development in the Los Angeles Scientific Laboratory, under the sponsorship of the Department of Energy. He has gone very seriously into all facets of it, and he is one of the few people who will give you numbers.

The top line on table 1-1 is predicted on the assumption that you can have semi-mass production. Of course, you cannot have mass installation of solar heating. This is always a handmade job, and you can't do very much to lower the

**Table 1-1**
**Cost of Solar Installations**
*($ per ft²)*

|  | Panels | Total |
|---|---|---|
| Los Alamos design | 6-7 | 12-15 |
| Professional design | 8-15 | 20-30 |
| Typical area needed | 400 ft² for 50% solar | |

price. The solar panels, on these assumptions, may cost $6 to $7, while the total installation cost is about $12 to $15 per square foot. The commercial design cost is listed in the second line; it is about twice as much. Even higher figures like $40 have been mentioned, but commercial prices also will come down as more units are produced.

In order to heat a house by solar panels, deriving about half your heat from them, you need a panel area roughly one-third the floor area of the house. So if the house has 1200 square feet, your panels are about 400 square feet.

The question is where can you use solar energy? Figure 1-2 again has been made by Dr. Balcomb. He has investigated the temperature and the sunshine in many places in the United States, and deduced from this how much solar heat you can obtain per square foot of panel if you want 50 percent of the heat for your house to be solar. (If you want 100 percent, solar heat becomes prohibitively expensive—at least on the present designs.)

You see in figure 1-2 some contour lines which show the contribution of solar heat. The contribution is especially high in the Rocky Mountains, where you can get over 150 units; at the coast of California, maybe even more. In Ithaca, New York, you can get only 80 units, and anybody who has ever lived in Ithaca knows why that is so: you don't have especially cold weather, but you have rather little sunshine in winter.

On the other hand, if you go down to Florida, there is lots of sunshine, but you hardly ever need any heat. Therefore, the contour line of 80 is also found in northern Florida, and the useful solar heat goes even lower in southern Florida.

So, you have to get a combination of large heating needs and large amounts of sunshine in winter. Otherwise, solar heat won't work very well.

The low numbers are about 80, and the high numbers are about 150 (see table 1-2). To this you may add some 40 or so units for water heating. Hot water is a much more favorable situation because you need hot water in summer as well as in winter, and even in Ithaca the sun does shine in summer.

So, adding these, you get a total supply of about 120 units in the bad areas and a little over 200 in the good areas. The units here are thousands of Btu's per square foot per year.

Now, let's calculate how much solar heat will cost. The ground rules on this

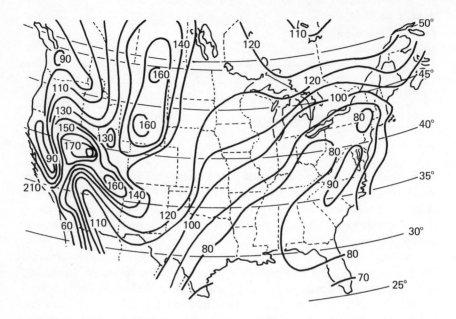

Courtesy: J. Douglas Balcomb and James C. Hedstrom, Los Alamos Scientific Laboratory.
**Figure 1-2.** Annual Solar Gain (thousands of Btu per year per square foot).

are as follows (table 1-3): you pay about 9 percent interest on the mortgage which you take out in order to get solar heat installed. You are in the 30 percent income tax bracket, which is perhaps about average. There are no property taxes on solar installations, a very important point. If you put property taxes on this, then solar heat is nearly dead. On the other hand, there are no subsidies. It costs you 2 percent maintenance per year, and you have either 7 or 20 years' life for the installation.

Depending on assumptions, you get very different costs per million Btu's as is shown in table 1-4. (M in the tables always means million, not thousand.) Under the most favorable conditions, you pay about $8 per million Btu's, and that means Los Alamos design, 20 years' life, the best area of the country for installation, and, in table 1-5, $8 per M Btu is a perfectly reasonable cost.

Under the worst conditions—7 years' life, Ithaca, New York, and the commercial cost of $30 per square foot—then it costs $50 per million Btu's, which is outrageous. (Please disregard the last line of table 1-5 for the moment.)

Table 1-5 is the most important one. Here I compare the cost per million Btu's from several sources of heat. In the case of solar (table 1-4), it is between $8 and $50. If solar is only used for water heating, you are a lot better off because solar is available summer and winter, and in that case, the cost is between $4 and $10.

**Table 1-2**
**Heat per Square Foot per Year**
*(thousands of Btu)*

| | |
|---|---|
| Best areas | 160 |
| Worst areas | 80 |
| Water heating | 40-50 |

Fifty percent of heat is assumed to be solar.

In most parts of the country today, natural gas costs $3 per 1000 cubic feet, which is the same as 1 million Btu's. So, if you have natural gas heating, keep it, don't change to solar heating. But many new houses cannot obtain natural gas because it is in scarce supply.

To make the gas synthetically, it will cost $6 to $7, which is based on the assumption that at the factory synthetic gas will cost between $3 and $4. Some people give good reasons why we should use the lower part of the range, because we will make gas in the places where it is cheaper, rather than where it is most expensive.

Fuel oil at 45 cents per gallon costs a little more than natural gas, but this price will go up steadily. Electrical resistance heating is hopeless. The figure given here is not based on the present price of electricity, but it assumes $.05 per kilowatthour, which is the price which you can anticipate when the power stations which are now being built come into operation. So it is the future price of electricity, not the present price which is much lower.

So, on this future price, which you have to be prepared for, electrical resistance heating is prohibitive, except that electrical resistance heating, more than almost any other method, lends itself to saving by heating only part of the house. I have lived in England for one winter. Electrical resistance heating was the mainstay of our heating. We heated only the room in which we were actually sitting, and the rest of the rooms were left cold. If you do that, then the cost of electrical resistance heating can be greatly decreased.

There is, however, another way to use electricity, and that is in heat pumps. In most weather conditions, a heat pump will give you about twice as much heat as the energy value of the electricity which you consume. In good circumstances,

**Table 1-3**
**Ground Rules for Cost Calculation**

| |
|---|
| 9% interest |
| 30% (marginal) income tax |
| No other taxes or subsidies |
| 2% maintenance per year |
| 7 or 20 years' life |
| 120 to 210 kBtu/(ft$^2$)(yr.) |

**Table 1-4**
**Cost of Solar**
*($ per MBtu)*

| Cost per Ft² | KBtu (Ft²)(Yr) | Life | |
|---|---|---|---|
| | | *20 Years* | *7 Years* |
| 15 | 210 | 7.90 | 14 |
| 30 | 120 | 27 | 50 |
| 10 | 300 | 3.70 | 7 |

at 40 to 50°F outside temperature, it will give you three times as much heat. However, if you go to very low temperatures, like 0°F, the factor goes down to about 1.

So, heat pumps will cost you about half as much as resistance heating, which means $7. Please disregard again the final line in table 1-5.

Now, the heating by gas and oil has to be corrected, because most furnaces are not fully efficient and use maybe 70 percent of the heat value of the gas or oil. If you apply this correction, then the natural gas goes to $4, the oil to $5, and the synthetic gas to $10. At this level you can see that, compared to synthetic gas, solar heating is competitive in the best areas of the country, if it is made in the cheap mass-production way of the Los Alamos Laboratory.

For water heating, solar is competitive almost everywhere today. (It is not competitive, of course, with natural gas, except possibly for water heating in some cases.)

This is the status of the economy of solar heating. If you take into account the effect of inflation (table 1-6), then solar heating looks better. Namely, take the present price of gas, assume it increases 10 percent per year in constant dollars until it reaches the value for synthetic gas, and assume inflation of 5 percent per year, which is mild. Assume maybe 8 percent interest on your money, minus tax. Then, to get 1 million Btu's for 20 years, by using gas, you

**Table 1-5**
**Cost per MBtu**

| | | *70 Percent Efficiency* |
|---|---|---|
| Solar | 8-50 | |
| Solar, water heating | 4-10 | |
| Natural gas | 3 | 4 |
| Synthetic gas | 6-7 | 8-10 |
| Oil (45¢/gal) | 3.3 | 5 |
| Electric resistance | 15 | |
| Electric heat pumps | 7 | |
| Electric off hours plus storage | 5-6 | |

**Table 1-6**
**Effect of Inflation**

---

Assume:  Gas cost: $3 per MBtu increasing 10% per year, in real dollars to $7
  Inflation 5% per year, interest 8% minus 30% tax
    70% heating efficiency

To get one MBtu for 20 years in gas, must invest $132

For solar energy,  best assumptions: $ 90
                   worst assumptions: $350

---

must invest today $130. But if you get yourself a solar installation, you only need to invest $90 at the best places in the country, which is good for solar energy. But in the worst places, and with a short life of your solar installation, the economics are still very bad.

Now, I shall mention another way of heating by using electricity. This has been proposed by Asbury and Muller in *Science,* February 4, 1977: use electricity to heat water in the off hours, that is, in the dead hours of the night, from about midnight to 6 a.m. During these hours, the demand on electricity is very low everywhere. During these hours, therefore, the electric power plants are underused. These plants have already been paid for; they must be there anyway to provide your electricity at 6 p.m. The distribution system also must be there anyway. So the only extra cost for delivering electricity in the off hours is the cost of the fuel.

To make use of the off-hour electricity, however, you must have in your house a storage system similar to, but somewhat simpler than, the storage system that is necessary with solar energy—to tide yourself over cloudy days. For solar, you must have a storage system which keeps the heat for several days. With the off-hour electricity, on the other hand, you need to tide over only 24 hours. So the storage system can be simpler.

Take, however, the full storage and pumping system which has been provided for solar energy, and then charge yourself for electricity only the cost of fuel, which is about 1 cent per kilowatthour instead of 5 cents. Then this off-hour electricity plus storage will give you 1 million Btu's for $5 to $6 (as indicated on the last line of table 1-5). Therefore, if we get sensible pricing of electricity in the off hours, which European countries have had for fifty years, then this type of heating is cost-competitive with any other type of heating which is likely to be available in the future.

Now, there is another way of using solar energy, and that is in large, central stations (table 1-7). This has been worked on for a long time by the Department of Energy (and its predecessor). The favorite method is to have a so-called power tower which receives solar heat from a field of perhaps 100,000 mirrors deployed around it, which reflect sunlight onto the tower. On the tower is a boiler which heats water by means of this reflected sunlight. The estimate of the

**Table 1-7**
**Cost of Large Power Units**
*[$ per kilowatt (electric)]*

| | |
|---|---|
| Solar power tower | 2500 |
| Photovoltaic | More |
| Satellite | Still more |
| Ocean thermal | 1300 |
| Geothermal | 1300 |
| Nuclear fission | 600 |
| Coal | 500 |

cost of such an installation, with all mass-production savings taken into account, has been made by the Sandia Laboratory in Albuquerque, New Mexico, which is the center of development of this means of heating, at $2500 per kilowatt. The present price under the same accounting, namely no interest and no inflation during construction, for a nuclear fission power station is $600, for a coal power station $500.

So, the solar power tower is hopelessly out of range, and this is realized by the people who actually work on it. Photovoltaic generation of electricity or satellites beaming down microwaves cost still more, but any numbers for these are pure guesses.

Use of ocean thermal currents may cost about half as much, with luck. Geothermal also about half as much, and by that I mean dry rock geothermal, not the use of hot water from the earth, which may be cheaper but not widely available. Wind power, I learned from an expert, will cost also about $1300.

Another use of solar energy is to get biomass. I am very much in favor of use of one type of biomass, namely the collection of waste from agriculture, forests, and urban wastes and transformation of this into methane gas, that is, substitute natural gas. To do this is at present still more expensive than the use of coal for the same purpose because of the cost of collection of the wastes. But it is not much more expensive, and it is a resource which we ought to use.

There is another kind of biomass which has been talked about, namely actually growing trees in order to convert them into synthetic fuel. In my opinion, this is something we should not contemplate, just from the standpoint of land use.

I have compared the amount of energy you can get from some acre of land out west, from which you can get surface coal, with the amount of energy you can get from an acre growing trees. I assume that the coal land is out of commission for ten years, which is more than is necessary, and is then restored. Then you can get from some good coal land out west 500 times as much energy as you can get under the best conditions by growing trees.

Finally, one other type of alternate energy, namely fusion, should be considered. The first point about fusion is that its feasibility has not been established yet. Everybody in the fusion community now is very optimistic that feasibility of fusion can be established about the mid-1980s. I agree with the fusion people that there's more than 50 percent probability that this will actually become true. Once you have fusion, you have the advantage of smaller radioactivity than with fission power.

However, the economics is very doubtful. Even the people who work on it and are advocates say that the price of a fusion power plant will be higher than of a fission power plant, and I assume it will be about twice as much for given power.

Everybody agrees that even after fusion gets established, it will take a decade or more before it can be engineered, and it will take more decades before it can be deployed in any useful way so that you can get an appreciable amount of energy from it.

Therefore, if we talk about energy before the year 2000, fusion is out; but after the year 2000, that is something different. Fusion may very well be in.

The most important use of fusion is that it can be used to make fissile materials—but that belongs into our third session and not into the first.

Therefore, to conclude with the beginning, namely, some alternate sources like solar heating are close to being realizable. But the large-scale use of solar heat is far beyond the presently foreseeable, and the same is true of fusion. So we come back to the three mainstays: conservation, coal, and nuclear.

**Part II
Realities of Energy
Conservation**

# 2 Public Policy Issues Relating to Industrial Cogeneration

*Robert Williams*

One of the most remarkable and hopeful impacts of the oil embargo of 1973, and the subsequent heightened general awareness of energy issues, is the creative ferment relating to new and renovated technologies for improving the efficiency of energy use. There are considerably greater opportunities for sustaining American life-styles with less energy than most energy analysts believed was possible a couple of years ago. This statement is valid not only in the narrow, technical sense but also in the broader, economic sense as well. For a wide range of conservation options, it will be less costly to the nation to save a barrel of oil per day of energy than to produce a new one.

Despite the growing body of knowledge relating to such opportunities, however, the United States' economy is only slowly adopting conservation measures. In part, this slow pace reflects the facts that considerable time is required both to turn over the existing energy-inefficient capital stock of the economy and to build up industries for the new conservation technology. But a more fundamental problem is that in order to capture the potential for conservation, new and rather unfamiliar institutional arrangements will have to evolve. Many existing institutional arrangements do not foster the rational, economic decision making appropriate for the era of rising energy prices.

This concept is illustrated by a particular conservation strategy—the cogeneration of process heat and electricity. Various technologies could be utilized for cogeneration. The remarks here will deal with two technologies: the steam Rankine cycle with back-pressure turbine and the gas turbine used with a waste heat boiler. The remarks will be restricted to cogeneration configurations at industrial sites, where electricity is produced as a by-product of industrial process steam. These systems are compared to the conventional situation where electricity is produced at central station power plants and process steam is produced separately at industrial sites (figure 2-1).

In cogeneration large fuel savings are possible. When electricity is produced as a by-product of a process steam, only about half as much fuel, beyond what is needed to produce steam, is required to produce a kilowatthour compared to what is required at a conventional central station power plant. The potential fuel savings in the overall energy economy can be quite large both because of this large specific savings and because industrial process steam accounts for about 13 percent of total United States energy use—about as much as the automobile. Cogeneration ranks with space heating and the automobile as an area where

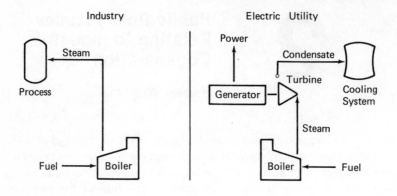

Cogeneration Plants

Back-Pressure Steam Turbine

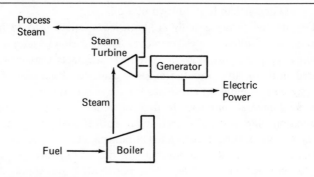

Gas Turbine—Waste Heat Boiler

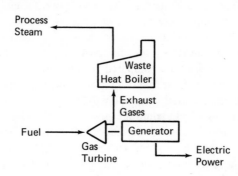

Source: U.S. Federal Energy Agency.

**Figure 2-1.** The Potential for Development in Six Major Industries by 1985.

potential energy savings at the national level could amount to millions of barrels of oil per day.

This fuel savings is often accompanied by a capital savings as well. This possibility is actually quite remarkable because industrial cogeneration systems are often only 1 to 10 percent as large as central station power plants. One would think that such units would suffer from diseconomies of scale. One way capital-cost savings would arise to offset scale diseconomies is through the sharing of capital expenses between steam and electricity generating activities. Also, for some cogeneration systems, such as those involving gas turbines, unit capital costs are inherently much lower than costs for central station power generation.

These fuel and capital savings possibilities are often reflected in favorable electricity costs for industrial customers. This result is illustrated in table 2-1 for large and small oil-fired gas-turbine cogeneration systems and coal-fired steam-turbine systems. These costs are calculated for a rate of return on investment appropriate for electric utilities so as to make meaningful a comparison of cogeneration power costs with the price of electricity today and the cost of electricity from new central station power plants. The lower cost shown for a given cogeneration unit would arise whenever capital costs can be shared between process steam and electricity, whereas the larger cost would arise whenever the cost of process steam equipment must be charged to power. The important point to note here is that over a wide range of circumstances, the cost of cogenerated electricity to industrial customers would be less than the cost of electricity from a new nuclear power plant. But, at the same time, the cost, in most cases, would be greater than the average cost of utility electricity today.

This situation arises, of course, only on the average. In regions such as the Northeast and the Midatlantic states, cogeneration is often economical at today's electricity prices. But wherever industrial electricity prices are comparable to or less than the average United States industrial electricity price, cogeneration often would not be viewed by a potential cogenerator as being as economically attractive as it truly is, simply because this potential cogenerator is shielded by the present utility rate structure from the full impact of replacement costs for electricity. Electricity rates today are set by rolling in the high cost of new plants with the much lower costs of electricity from existing units. Thus prices will rise only slowly to the replacement-cost level. Until acceptable institutional mechanisms are developed that sensitize potential industrial cogenerators to replacement costs for electricity, the full economic potential for cogeneration will not be realized.

One important consideration relating to cogeneration is the overall fuel savings potential to the nation arising from the displacement of central station power generation. The potential fuel savings varies substantially from technology to technology, as shown in table 2-2. Here projections of the national cogeneration potential are given for alternative technologies in the year 2000. Note from this graph that the fuel savings potential is low for steam-turbine systems compared with the savings potential with gas turbines and diesels. These

**Table 2-1**
**Levelized Cost of Electricity to Industrial Customers**
*(mills/kWh, 1976 dollars)*

| | Central Station Power | | Cogenerated Power | | | |
| --- | --- | --- | --- | --- | --- | --- |
| | Average Cost of Utility Electricity in 1976 | Cost of Electricity from New Nuclear Plant | Oil-Fired Gas Turbine | | Coal-Fired Steam Turbine | |
| | | | 10 MW (electric) | 80 MW (electric) | 5 MW (electric) | 30 MW (electric) |
| Capital | | 14.9 | 5.4- 7.5 | 3.2- 4.6 | 12.2-23.6 | 7.0-13.7 |
| Fuel | | 5.2 | 15.4 | 15.4 | 4.3 | 4.3 |
| O & M | | 2.3 | 4.6 | 4.6 | 3.0 | 3.0 |
| Bus-bar Costs | 12.4 | 22.4 | 25.4-27.5 | 23.2-24.6 | 19.5-30.9 | 14.3-21.0 |
| T&D Losses | 0.6 | 1.2 | | | | |
| T&D Costs | 7.7 | 7.7 | | | | |
| Average Cost of Delivered Electricity | 20.7 | 31.3 | 26.0-27.9 | 24.0-25.3 | 23.7-34.0 | 18.6-24.7 |

**Table 2-2**

**Cogeneration Potential for the Year 2000 if Half of Industrial Steam Is Associated with Alternative Cogeneration Technologies**

|  | Fuel Savings | Generating Capacity |
|---|---|---|
|  | $(10^6$ barrels of oil per day) | $(10^6$ KW) |
| Steam Turbine | 1 | 70 |
| Gas Turbine | 3.5 | 280 |
| Diesel | 5 | 560 |

differences can be very significant at the national level. The cogeneration capacity, expressed here as the equivalent base-load central station generating capacity, could range from a low of about 70 gigawatts with steam turbines to over 500 gigawatts with diesels. For comparison, today's electricity consumption could be provided with the equivalent of about 350 gigawatts of base-load capacity.

Both technical and institutional factors which limit the technologies offer greater energy savings, however. Steam-turbine-based cogeneration is the favored technology today for two reasons. First, it is the more familiar cogeneration technology. Second, it is the only one that can be used with coal today. Most gas-turbine and diesel systems today burn high-quality fuels instead. However, it is also possible to burn residual oil in these systems, and further technological advances are expected in this area over the next several years. Moreover, with appropriate R&D it should be possible to bring to commercialization within a decade one or more high fuel saving cogeneration technologies based on coal. The possibilities include the indirectly fired gas turbines with atmospheric pressure fluidized bed combustors, directly fired gas turbines with pressurized fluidized bed combustion, coal-fired diesels, and even Stirling engines.

A cursory review of the technological possibilities suggests that one should not assume that the optimal technologies for cogeneration in the long run are to be found among the technologies with which we have had the most experience to date in power generation.

One of the most intriguing aspects of this area of endeavor is that there are really vast opportunities for technological innovation compared with the prospects for innovation in central station power generation. As shown in table 2-1, the cost of electricity from a new central station power plant is considerably greater than the average price. There are no obvious courses open for reducing these costs. This situation represents a dramatic reversal of historical trends. As figure 2-2 shows, the price of utility electricity declined dramatically in the past. While residential consumers obtained the greatest benefits from cost-cutting innovations, the benefits to industrial customers were far from trivial. The real price of electricity to industrial customers was reduced by half between 1940 and 1950 and again between 1950 and 1970. From the turn of the century

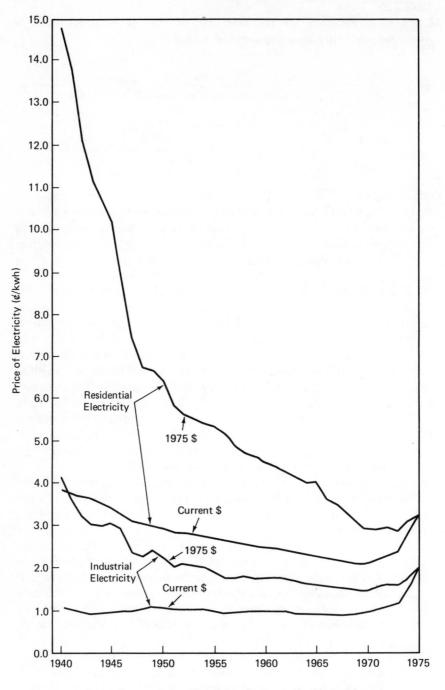

**Figure 2-2.** Average Electricity Prices in the United States.

up until about 1960, much of the cost reduction was associated with efficiency improvements (see figure 2-3). But, subsequently, there have been no further improvements in efficiency. After about 1960 much of the cost cutting was achieved through scale economies, as figure 2-4 shows. But most analysts agree that opportunities for achieving further scale economies are limited, and some feel that the optimum size in power generation has already been exceeded.

In sharp contrast to this situation, one can see distinct possibilities for cost-cutting innovations in cogeneration. One of the more hopeful prospects is the potential for low-cost electricity from cogeneration systems based on pressurized fluidized-bed combustion (PFBC), gas turbines, and waste heat boilers. The PFBC units should be quite small relative to conventional pulverized coal boilers (see figure 2-5). This reduced size could well be reflected in lower capital costs, both because of reduced material requirements for fabrication and because of the potential for shop fabrication instead of the more costly field construction. A real but still uncertain possibility is that the cost of electricity produced via such cogeneration systems would be not only less than the cost of electricity from alternative new sources of power but also less than the average electricity price today.

There are significant institutional obstacles to widespread implementation of high fuel-saving cogeneration technologies. But because of fairly strong economic incentives these obstacles are not insurmountable. The principal institutional difficulty arises because with these systems often more electricity is produced than can be consumed at an industrial site. Therefore there must be a market for the excess electricity. It would be logical for a cogenerating industry to sell electricity to the utility. Unfortunately, utilities have tended to discourage such activity because they have usually been willing to pay industrial firms much less than what these firms would need to realize an adequate return on their investments. Thus, industry has focused its attention mainly on those

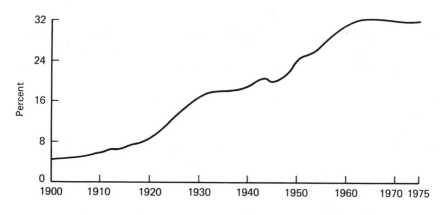

**Figure 2-3.** Average Steam Electric Power Plant Efficiency.

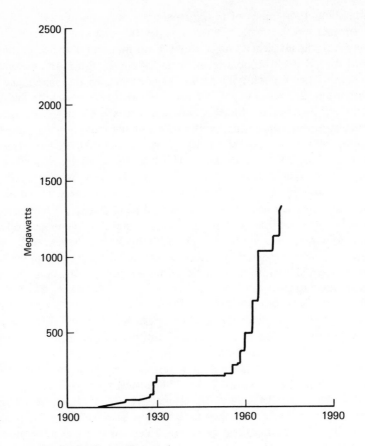

Source: U.S. Federal Power Commission.
**Figure 2-4.**  Largest Steam Electric Turbine Generators in Service.

cogeneration technologies that avoid the excess electricity problem. Unless this obstacle to the sale of excess power can be overcome, the nation will be denied the considerable fuel savings associated with the very efficient cogeneration technologies.

There are many interesting possibilities for overcoming the problem, but they represent largely unfamiliar practices. The strategy most discussed today is to adopt regulations that require utilities to pay a fair price for this excess electricity—perhaps a price equal to what it would cost the utility to generate the electricity itself from the least costly alternative new source. But there are many other possibilities as well.

As long as cogenerated electricity is less costly than alternative new sources of electricity but more costly than the average utility electricity price, the arrangement that holds the most promise is utility ownership of cogeneration units at industrial sites. This arrangement would facilitate an evaluation of

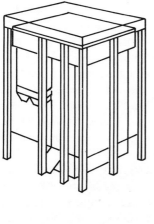

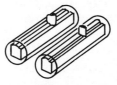

0    50    100    Scale Feet

Note: The unit at the left is a conventional pulverized fuel boiler. The unit at the right is a pressurized fluidized-bed combustion boiler with a bed pressure of 16 atm.

**Figure 2-5.** Relative Sizes for Alternative Coal-Fired Boiler Designs for a Common Boiler Capacity of 660 megawatts.

cogeneration projects based on long-run incremental costs, and it would lead to more projects being judged economical than with industrial ownership, since utilities would usually require a lower rate of return on their investments. This latter advantage would also mean lower rates for consumers generally.

The option of utility ownership should allay many of the utility's concerns about cogeneration. Utility ownership brings cogeneration into the rate base, thereby providing a utility incentive to cogenerate. Utility ownership also provides a straightforward means by which utilities could maintain dispatching control over the electric power entering the grid, to ensure system stability and security.

One concern that many industrial firms might have about utility ownership is that the rate charged by the utility for power might change in the future in unpredictable ways. For example, the implementation of lifeline rates for low-income residential customers might be offset by a general increase in the rate for industrial customers. No such problem would arise with industrial ownership. It may be desirable to protect the industrial firm against this prospect whenever cogeneration is competitive with alternative new sources of power. Such protection could be provided through some degree of deregulation of the rate agreement between the utility and the industrial firm(s) involved in a cogeneration project.

If new technologies emerge that enable cogenerated electricity to be produced at less cost than today's average electricity price, then very interesting new institutional arrangements might be feasible. A whole new industry might be developed for owning and managing industrial cogeneration facilities. It might be easier for new industries to be organized than for today's utilities to be reorganized to accommodate the decentralized operations required for cogeneration. Because such new industries would be in the business of selling power, they would probably be more interested in electricity export than would the industrial firm being served—if it were profitable. While these cogeneration firms would manage dispersed operations at industrial sites, they could be more centrally organized so as to share technical services among many sites. As new technology becomes available, such industries might evolve as affiliates of vendors so as to facilitate the introduction and servicing of new technology.

Whatever ownership arrangements emerge as the most viable for cogeneration, one factor is clear: some degree of deregulation of power generation is probably called for. Policy changes in the direction of deregulation probably would have a much greater impact on the development of cogeneration than the investment tax credits and other incentives usually proposed to foster cogeneration. One of the principal original purposes of regulation was the recognition that scale economies in generation made inefficient a situation with each of many firms capturing a small share of the market for electricity. But small-scale cogeneration can be cost-effective today, so that the rationale for regulation breaks down. Introducing competition into power generation would also be a stimulus to technological innovation, which is key to the long-term success of cogeneration.

These considerations point to new directions for research, development, and demonstration. We usually think of RD&D in terms of hardware. And, indeed, the hardware RD&D will continue to be very important for cogeneration because the possibilities for innovation are very great. However, an important new dimension of the overall RD&D effort should be to carry out, with present technology experiments with pricing policies, various degrees of deregulation and alternative ownership arrangements in institutional demonstration projects over the next five to ten years as we wait for new technology. The focus here should be on those arrangements that involve the high fuel-saving cogeneration technologies for which there would be a strong coupling of cogeneration units to the utility grid. At the same time, perhaps, the highest priority for energy policy research relating to cogeneration should be to examine carefully the benefits and drawbacks of various degrees of deregulation of power production via cogeneration.

If these activities are successful, then cogeneration could become the major source of new base-load electric generating capacity in the United States in the last ten to fifteen years of this century as the transition is made to an industrial energy economy where process heat is provided in the much more efficient manner required for the new era of high-priced energy.

# Major Factors of Conservation in Industry

## Macauley Whiting

A report by the Committee on Nuclear and Alternative Energy Strategies, which the National Academy of Sciences is sponsoring for the Department of Energy, is looking at energy supply and demand in the year 2010. The study is particularly concerned with the need for breeder-reactor development, which, of course, would not have any major influence on United States energy use until about 2010.

The demand/conservation panel of this study has taken the first comprehensive look at future United States demand for energy, including long-term projections for these 35 years till 2010.

The industry-sector group of this panel identified the likelihood that the energy intensity of industry will be down 25 percent or more by 2010. Energy intensity is the energy used per unit of output—per ton of steel or per pound of petrochemicals. The panel concluded that the energy required to maintain a certain standard of living for the United States at any time, any particular time, is primarily a function of the stock of energy-using equipment and buildings, that is, industrial equipment, the transportation equipment, the household equipment, and the buildings.

By world standards, the United States stock is not very efficient. Sweden, for instance, uses approximately onehalf the energy per person used in the United States, for an equivalent standard of living and an equivalent standard of industry, and that's just a very rough number.

The conclusion that you reach then from all this is that the United States can satisfy the full public needs and wants with considerably less energy per person than is being used today. This would be done by building a more energy-efficient stock of buildings and equipment. These will be more expensive than the present stock, but not very much more expensive.

The overall cost of living will go down, rather than up, with conservation in the face of the higher energy prices, which we all know are coming.

The achievement of energy efficiency will take a long time. We must replace the present stock of buildings and equipment. For automobiles, it takes about 10 years to turn over the stock; for industry, industrial equipment, perhaps 30 years; for buildings, maybe 60 years. So conservation cannot be accomplished in the short term.

The same thing can be said in another way. There are a number of countervailing forces at work. In the United States we're going to have more people, we're going to have a higher standard of living, and those things will work toward higher energy consumption. The change of equipment and buildings will work toward lowering energy consumption.

It has been suggested that this can lead to a standoff and that no more energy will be required in the year 2010 than is used by the country today, while at the same time we are achieving a much higher standard of living.

Now, let's go to industry. The fundamental parameter of energy used is energy intensity, and the energy intensity for industry should drop 25 percent or more by the year 2010.

Most people hope that industrial production will grow at a rate more than equivalent to the rate of conservation. Thus total energy use will increase, but at a much slower rate than it has in the past.

Now, for energy accounting purposes, there are two classes of energy users in industry. First are industries which consume energy to produce nonenergy products and services. Second, there is that special segment of industry whose product is energy, but energy in a more useful form than occurs naturally. In this segment are, for example, oil refineries and electric utilities.

The major energy users in the general category are listed below. These few industries account for more than 70 percent of all the energy use by consuming industry.

| | |
|---|---|
| Agriculture | Food |
| Aluminum | Glass |
| Cement | Iron and Steel |
| Chemicals | Paper |
| Construction | |

The energy-producing industries are shown in the second list. They were not the subject of study by this demand/conservation panel, but it's worth noting that they will consume an increasing share of the energy of the United States in the future. The primary energy forms to be used—coal and nuclear—are less suited to the average consumer without processing by these energy-producing industries.

Oil and gas extraction

Oil refining

Coal mining

Synthetic fuels

Electric power generation

In reviewing the list of consuming industries, it is evident that, with the exception of food and agriculture, they are all heavy industry, processing basic materials.

Now, table 3-1 shows the energy consumption by these industries—past, present, and future. The future projections are a couple from over a dozen made on different bases by this panel. Table 3-1 shows the total for all industry, and that in turn is related to comparable total United States usage of energy. It is clear that the share of energy taken by industry grows over time. This is partially based on the increasing popularity of electricity and other processed forms of energy, so that these energy-processing industries are taking a very much increasing share of the energy slate. In addition, the increasing use by industry has been based on the increasing share of the GNP that is to be taken by industrial products.

Another facet of industrial energy consumption is the diverse applications to which energy flows and the complex network of paths by which it gets there (figure 3-1). The primary energy sources—fuels—are shown at the left-hand side of this figure. The end-use applications in industry are shown on the right-hand side. Optimizing the energy flows in this network is a challenging task that demands a high degree of skill and experience. Perhaps examination of this figure will help you understand why industry prefers the pricing mechanism as motivation to conserve, as compared with rigid regulation. It is all but impossible to devise a reasonable and equitable scheme of government regulation for such a complex network of flows and energy.

Figure 3-2 plots the good history of overall industrial energy intensity. It reveals that the energy intensity of United States industry has been dropping from 1920 to the present. The demand/conservation panel study projects a continuing decline at the same rate. Again, there are higher prices expected

**Table 3-1**
**Energy Use by Industry**
*(quads)*

|  |  |  | 2010 | |
|---|---|---|---|---|
|  | *1950* | *1975* | *Low Projection* | *High Projection* |
| Energy-Producing Industry | 5.5 | 16.0 | 35.0 | 60.0 |
| Energy-Consuming Industry | 11.0 | 20.0 | 40.0 | 55.0 |
|    Basic Materials | – | 13.0 | 31.0 | 43.0 |
|    Other | – | 7.0 | 9.0 | 12.0 |
| Total Industry | 16.5 | 36.0 | 75.0 | 115.0 |
| Total United States | 34.0 | 71.0 | 110.0 | 160.0 |
| Feedstocks | 0.5 | 3.0 | 7.5 | 10.5 |

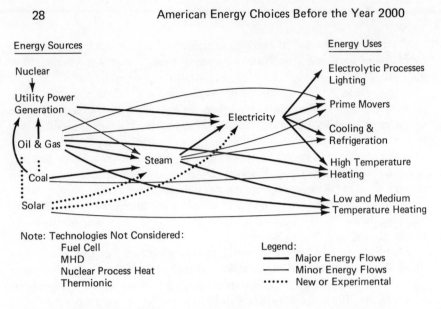

Note: Technologies not considered: fuel cell, MHD, nuclear process heat, thermionic.

**Figure 3-1.** Alternative Paths for Industrial Energy Supply.

which force a higher rate of decline, but there are other factors that will make conservation more difficult, that counter the tendency to improve.

The other energy-consuming sectors—transportation and buildings—have not shown similar improvement in the past, but they are expected to in the future.

These are the major actions that industry must take to conserve energy:

Replacement

Retrofit

Process innovation

Fuel conversion

Cogeneration

Industry will incorporate more energy-efficient equipment into its manufacturing facilities as growth occurs and obsolete items are replaced. In most areas, new energy-consuming industrial equipment is continually becoming more efficient. For instance, centrifugal compressors for gas have largely replaced reciprocating machinery for the large-volume movement of high-pressure gases. The shift has brought significant energy savings. Computer-aided calculations have been described as another practice by which new facilities are made more efficient than the old.

It's not unrealistic to expect that a new industrial facility will use

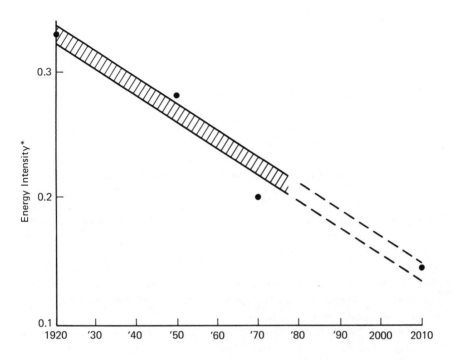

*Energy intensity = ratio of total manufacturing energy use to FRB production index.
**Figure 3-2.**  Industrial Energy Conservation Trend.

one-quarter less energy than the average of the existing plant. And if the new facility is replacing a very old plant, the savings can be considerably more than 25 percent.

Conservation by replacement is relatively painless when an old facility is being retired at its normal age. However, when the retirement is accelerated specifically to accomplish energy conservation, the justification for the required capital may be difficult.

In general, even at today's energy costs, a 25 percent saving in the fuel bill will not be sufficient to justify the early retirement of a major energy-consuming facility.

With respect to retrofitting, when the replacement is not economical and some gains can be achieved by retrofitting old facilities, more energy-efficient components can be introduced into old process trains. The total cost of doing this is, of course, less than replacement of the whole train, but the energy savings are also less. Short-term gains, and that is between now and 1980, in efficiency of industrial energy use will come primarily from replacement and retrofit. Based on data compiled in the voluntary industrial conservation program of the

Departments of Energy and of Commerce, a 15 percent improvement in energy efficiency appears to be obtainable by 1980.

More glamorous than the mundane business of replacement and retrofit is the introduction of major new industrial processes. A recent example of such a process is the float glass process used for glass for windows and similar applications. This process enabled roughly the same aggregate of plant, labor, and energy to produce three or four times as much glass as had been produced previously. The process has taken the glass industry by storm and resulted in very significant conservation. There are other important processes being worked on currently. A process for aluminum manufacture, which saves about 25 or 30 percent of the energy now used in manufacturing aluminum metal, is one that's received a great deal of publicity.

Then there is the conversion to alternate fuel, a shift away from oil and gas, primarily to the much more available coal. Steam generation, which represents about 40 percent of industrial fuel demand, is the major proposed application for conversion. There are considerable penalties that come with the utilization of coal. However, the penalties are relatively smaller with the use of solid fuel in large boilers as compared with small ones. The problems of solids handling, ash removal and disposal, and air pollution coming from coal can be solved much more economically on a larger scale.

What, then, is the bottom line of energy savings as a result of the above-mentioned industrial conservation actions? After factoring the energy costs of environmental protection, we expect to have a further reduction of 25 percent. A few other figures may be of interest. Natural gas, which presently constitutes about 40 percent of industrial fuel, will decrease in share to 15 percent, while coal increases from 25 percent up to 30 or 40 percent.

The use of replenishable fuels, wood waste, solar, etc., will not become important by 2000. The share of industrial energy provided by these sources is expected to increase from 5 percent today to something well under 10 percent.

# Citizen Participation in Conservation

*Shirley Sutton*

Project Pacesetter is a citizen energy education and conservation project involving the whole community. The idea of a Pacesetter-type program began with the national Americans for Energy Independence (AFEI) organization over a year ago. AFEI was concerned about the fact that almost 50 percent of the American public did not believe there was an energy problem; so they decided to try a major citizen education program. Allegheny County, Pennsylvania (which includes the city of Pittsburgh), was chosen as the location to try this pilot program.

An executive committee of leading Allegheny County citizens was chosen to direct the project. The committee was chaired by Richard Simmons, president of Allegheny Ludlum Steel, and Lloyd McBride, president of the United Steelworkers of America. Serving on the committee were Edgar Speer, chairman of the board of United States Steel corporation; Joseph Odorcich, vice president of United Steelworkers of America; Dr. Wesley Posvar, chancellor of the University of Pittsburgh; Susan Brandt, president of the league of women voters, Allegheny County Council; Ronald Davenport, dean of the School of Law, Duquesne University; and Commissioner Jim Flaherty, chairman of the Board of Commissioners of Allegheny County.

Each member of the executive committee headed a task force. The six task forces were designated to represent various segments of the community: education, labor, business, government, volunteers, and public interest groups.

The project was kicked off in the spring of 1977 with a community announcement that Allegheny County was chosen to try this type of project. Throughout the late spring and summer, each task force organized within its area of influence. Other county leaders were chosen to head subtask forces. These groups met, gathered ideas from people with expertise in their fields, and drew up a plan to be implemented in Allegheny County. The plans told of ways that each segment of the community could contribute to possible solutions to our energy dilemma. Suggestions were also offered for the county as a whole.

September 19, 1977 was Energy Day in Allegheny County. This was the day that the task forces announced to the community the results of all their planning. Each task force had come up with many suggestions, but was instructed to, and did, announce one major project that it would like to emphasize. The educational task force's primary recommendation was the development of an energy curriculum in the school systems of Allegheny

31

County. The labor task force recommended that management-labor energy committees be established to discuss energy-related matters. The government task force announced an "energy alert" system, designed to measure the amount of energy being used through the course of the year in Allegheny County. The volunteer task force announced a plan to build models of a home to show people how their homes could be insulated and otherwise weatherized. The public interest task force mentioned recycling as their major project, while the business task force suggested the development of energy management workshops, with all businesses designating an energy manager responsible for conservation.

Since September 19, the task forces and the Pacesetter staff have been working to implement, first, these primary recommendations and, after that, a number of the other recommendations—as many as we possibly can work out within our time constraints.

When AFEI drew up the Pacesetter plan, they did so with the idea that if Pacesetter was successful, similar plans might be started in many other urban areas in the country. If, through Pacesetter, more citizens in Allegheny County are convinced that the energy crisis is real, and if these same people are convinced that they should conserve in every way possible while solutions to the energy problem are being worked out, the organizers will consider an expansion of Pacesetter. This is the reason that *Pacesetter* was chosen as the name for the project, on the assumption that the project in Allegheny County will set the pace for the rest of the country in energy conservation activities.

How will the success of Pacesetter be determined? From the beginning, it was decided that there would have to be a method of evaluating the program, and plans were made to have this evaluation done by one of the local universities. As the project began and moved forward, interest was shown in the project by the then Federal Energy Administration. In September 1977 the Department of Energy announced that the University of Pittsburgh would receive a grant to monitor Pacesetter.

In addition to reaching out to the people through their organizations such as clubs, professional groups, service groups, labor unions, etc., a mass media campaign was conducted in the early phases of the project. This effort developed awareness of Pacesetter in the community, and the project continues to enjoy good coverage.

Is it possible at this point to evaluate how the project is going at the half-way point? At the outset, expectations may have been too high. Plans were extensive and perhaps unrealistic. It is a monumental job to organize a community to accomplish all the many goals that had been discussed in the beginning of the original time frame. Nevertheless, it is likely that all the major recommendations of the task forces will be achieved. As a pilot program, Pacesetter has involved a learning experience. Many of the programs and plans that were tried and found unfeasible would not be tried at another time in another city, thus saving the expenditure of time and energy.

Many people have said, as expected, that they don't believe in the "crisis," but many more have given time, money, materials, and "in kind" services. There were the usual groups or individuals who didn't carry out the job assigned to them—and others who have done an excellent job with their projects. Very few people have been unwilling to do something as part of Pacesetter.

This project has been a tremendous challenge. It has been a great opportunity to work with diverse segments of the community. People with opposing opinions have talked over energy problems together and are working together. It is not known yet whether many opinions are changing—or if larger numbers of people are conserving energy—but the project appears to be reaching great numbers of people. The evaluation should provide some concrete results on the basis of which it can be reported that people working voluntarily together in a local structure can make a difference.

# 5

# Impact of Conservation on Low- and Fixed-Income Americans

*Clarke Watson*

In August 1977, the Technology Assessment Board of the United States Congress, chaired by Senator Edward M. Kennedy, released a report authored by the staff of the Office of Technology Assessment to provide "a balanced and impartial analysis of the Administration's proposed National Energy Plan." The *Executive Summary* of the report stated:

> The National Energy Plan's assessment of the world energy crisis is accurate. The problems are complex and serious and there is little time for fashioning new policies to respond to them. If the United States acts now, it may be able to reassert control over its energy future and prevent serious economic, social, and environmental impacts. To postpone decisions to raise energy prices and reduce energy waste is to risk losing that control, which could mean severe hardships for all Americans within the next 10 years. The level of U.S. oil imports is the pressure gauge that will measure how well American policies are succeeding. If imports can be held close to the goals of the Plan, the United States and the rest of the world may well manage a relatively smooth and peaceful transition to sustainable energy resources. If not, the transition may be neither smooth nor peaceful.

> The National Energy Plan correctly acknowledges that energy problems exist on so many levels and in so many time frames that they must be addressed on a national scale. National security, economic stability, and other national interests are at stake. Decisions on energy must be made in consultation with State and local governments and with public and private interest groups, but the policies should reflect national concerns.

We agree with this assessment of the energy situation. At page 153 of the report, it is observed that "The National Energy Plan will impinge on many explicit and implicit social goals. The economic impact will vary by income class, region, and sector, posing equity questions that may require mitigating policies." It also notes that

> the share of income devoted to energy-related expenditures falls sharply as income rises. One estimate is that the lowest-income quartile spends more than 30 percent of its income directly or indirectly on energy, while the highest quartile spends about 10 percent. The Plan does not

address this issue in a substantive way. It promises "a reformed welfare system" and a "redesigned emergency assistance program" to help (p. 90), but these proposals may not go far enough to protect low-income families. Even the Plan's proposed per capita rebate of wellhead taxes will not necessarily assure equity because not all of the tax will be rebated to individuals (some will go to offset revenue losses from investment tax credits) and the tax will not be rebated progressively (Sec. 1403). The proposed welfare and emergency assistance programs may aid the poorest groups but those just above that level are likely to have the largest burden imposed upon them by the overall energy situation.

If there should be an added increment to inflation, as seems likely, or if the Plan should prove to adversely affect economic growth, lower-income groups will bear the brunt of this. The young may be affected by a further slackening in job opportunities, coupled with added inflation. The situation of the young, as affected by the Plan, is not addressed.

The poor, and particularly the rural poor who probably comprise most of the half of the lower-income group who own cars, will be hit most heavily by the increases in gasoline prices the Plan proposes. Not only do they spend a relatively larger proportion of their income on gasoline, they suffer from two other handicaps that would make it difficult to adjust to higher transportation costs. First, mass transit is not available for all essential travel, such as to work. Secondly, the poor generally cannot afford new, gas-economizing cars. They will be the purchasers in the second-hand market of "gas guzzlers" whose relative prices will fall as gas prices rise, bringing them within reach of lower-income groups. Thus, those who can afford new, fuel-efficient cars will be saving money on gasoline while the poor will be spending more on gasoline. No element in the Plan recognizes or offsets these possible inequities.

A comparable lack of capital will preclude lower-income homeowners from taking advantage of the tax-credit programs for residential insulation or solar energy units. They may not be able to meet "front end" costs and they may not be paying enough taxes to get the full tax credits proposed by the Plan.

One group of Americans who will not be able to benefit from residential energy conservation programs are tenants who pay for their fuel but who cannot be reimbursed for insulation expenses. Tenants who do not pay for the added cost of heating oil directly will do so indirectly through higher rents, but they are not likely to benefit from rebates on home-heating oil. The Plan's proposed increase in the federally financed weatherization program will help in sheltering the poor against higher fuel costs. However, the current program does not extend such help to renters. In addition, the level of funds available for insulation assistance may be too small. At present, there are approximately 9 million substandard homes in the United States, homes which for the most part are inhabited by the poor. The weatherization programs will handle only a small fraction of these structures.

Probably no plan could forsee and offset all inequities. What the Plan could include, however, and what is lacking, is a program to monitor its equity effects and those of the general energy situation and a mechanism for effectively proposing programs to redress inequities.

This is the crux of the problem of the administration's energy policy. Although there have been three subsequent messages from the President to the public, no mention has been made about this aspect of the energy policy so particularly vital to the well-being of black Americans.

Black Americans are not only disappointed that this area continues to be ignored, but are dismayed at President Carter's persistence in desiring to add enormous taxes to the market cost of energy. Past experience has shown that substantial contributions to the federal weal have not yielded an equal and substantial alleviation of the plight of black Americans. Grandiose programs such as The War on Poverty and Model Cities were dismal failures because they did not confront the root and continuing cause of disproportionate poverty which was and remains the result of racial discrimination.

Discrimination means exclusion from centers of economic power and influence. It is no small coincidence for example, that blacks are well represented in the highest offices of Labor, HUD, HEW, and Justice, but only nominally so in Treasury, Defense, Commerce, OMB, CEA, DOE, and CIA. All these agencies, as distinguished from Labor, HUD, HEW, and Justice, are major centers of enormous concentrated economic power and influence.

The same is true in the private sector where, except in rarest cases, the upper echelons of American enterprise continue to be lily-white. Thus, programs of quality education and busing and integrated housing and adequate representation before the courts come to naught since they fall far short of totally integrating the American fabric. In other words, while blacks have made significant social advances, attempts at economic parity have failed dismally.

The national energy policy which could provide a contrast to this historic dilemma instead seems bent on its perpetuation. The policy does not say, for example, that in each of the nonrenewable energy sources—coal, oil, shale, gas, and synthetic conversions—there is much room for specific economic development for blacks. The policy fails to take into consideration that in renewable resources—solar, wind, geothermal, OTEC, and fusion—specific guidelines should be set forth to ensure maximum economic opportunities for blacks. In essense, for black America the national energy plan is a reaffirmation and solidification of the status quo.

An energy scenario emphasizing an expanding supply of energy will allow the transition from fossil fuels to fusion and solar (and perhaps beyond), without the displacement of jobs and income that seem to be embodied in conservation proposals currently being articulated by government and groups in tune with nature but not in harmony with man. No enlightened civilization should reach for easy solutions, thought to be politically attractive, which jeopardize a third of its people—that third which can least endure further social and economic hardships.

# 6

# Economic Constraints on Federal Conservation Targets

*Arnold Moore*

The petroleum industry produces energy, but it also consumes very large amounts. Therefore it has been very active in conserving energy inputs as they have become more costly. In such a competitive industry there has been tremendous economic pressure to do so. As a result, energy consumed per unit of output has been reduced by about 12 percent since 1973. The industry believes that conservation is important and ought to be pursued, but that increased production must also be supported because what can be accomplished by conservation alone is limited. This is especially so if economic growth is to continue.

A discussion of the constraints on conservation involves four topics: First, what is meant by conservation? Second, what is the nature of the choices involved in a conservation-oriented policy? Third, is the United States wasteful of energy? And fourth, how much difference can conservation make?

To begin with the word *conservation* itself, in most discussions it is not defined. This gap accounts for much of conservation's appeal, since disagreements among its supporters are less apparent. Even when it is defined, definitions are generally empty or internally inconsistent.

There seems to be a definition of conservation implied by current arguments for it. This is, "Conservation is levels of consumption and production below those which would otherwise occur in response to current and projected market forces." This raises a couple of interesting questions. Why will response to market forces lead to excessive use of resources? In general, it seems that market prices and interest rates do lead to appropriate levels of consumption, though there are cases where this is not so. Is there a national security interest in stability of supply? Another question may relate to the effects of energy consumption on the environment. Both of these could be reasons why market forces would lead to "too much" consumption. But the large amount of environmental legislation on the books probably deals with the latter problem.

The national security aspect of energy policy seems very important, though it is little discussed. The planned Strategic Petroleum Reserve, containing up to 1 billion barrels of crude oil, is one appropriate response to this problem. Whether this reserve obviates the need for reductions in consumption beyond those arising from market forces has not been seriously discussed. Most important, the general question of why market forces are inappropriate, and to what extent, is little discussed in considerations of how much conservation is desirable.

A related question is, What determines market levels of consumption and production? Habit, technology, available resources, the existing capital stock, culture, psychology, law (including tax law), tastes, history, religion, climate, population density and mobility, and prices all affect consumption and production. Of these only prices—and to a lesser extent laws—can change quickly. To most people, conservation means changing these nonprice determinants of behavior, but this is a slow process. This explains why the President's energy plan has relied so heavily on market mechanisms via a complicated, and apparently unpopular, mix of taxes and subsidies to discourage consumption. Even this pseudo-market approach has its limits. On the consumer side, the gasoline tax proposed to reduce demand was defeated in Congress for what are called political reasons. On the industrial side, limits to conservation taxes are set by potential employment effects, competition from imports, and alternate demands for capital.

This discussion suggests that the nonprice forces determining the level of personal consumption are slow to change and the use of taxes to achieve conservation is limited by external market and political forces, along with others. Thus movement to levels of consumption very different from those implied by market forces is likely to be a slow and complex process.

As for the nature of the choices involved in energy conservation, the conservation argument goes like this: "Oil is a finite resource and we are running out. Every barrel we consume is a barrel less left for future consumption. As a result, we must consume less now." This is a proposal that we consume less now so that we or others can consume more in the future. There is nothing inherently wrong with this, of course, and I do not ask, as Mayor Daley did, What has posterity ever done for us? We should, however, consider our own and posterity's interests directly. Let's ask ourselves why less oil consumed now and more oil consumed later is a preferred pattern for the United States as a whole—or the world as a whole. Most of us expect posterity to be richer than we are. If so, we may ask, Why should we shift consumption forward to them? Conserving for *their* benefit seems like a kind of regressive tax on us, and we oppose regressive taxes.

People are rightly suspicious of a call for present sacrifice to avoid future danger. Guiccardini, the Florentine historian, four centuries ago wrote that

> Excessive forethought and too great solicitude for the future are often productive of misfortune; for the affairs of the world are subject to so many accidents that seldom do things turn out as even the wisest predicted; and whoever refuses to take advantage of present good from fear of future danger, provided the danger be not certain and near, often discovers to his annoyance and disgrace that he has lost opportunities full of profit and glory, from dread of dangers which have turned out to be wholly imaginary.

The American public understands this argument very well, and many remain unconvinced that "the danger is certain and near."

A conservation-oriented policy involves more than reducing consumption now to increase it later—a shift in timing. In addition, it involves use of other resources to substitute for oil and for energy. Thus, use of aluminum and fiber-glass permits insulation of buildings and reduces energy used to heat and cool them. However, it takes lots of energy and other resources to make storm windows and fiber-glass. In addition, aluminum is presumably drawn from a finite stock just like oil. Thus conservation involves trading depletion of one resource for another.

The nature of the choices involved in a conservation policy is extremely difficult. It involves intertemporal choices of a kind everyone finds difficult. Further, the human tendency to procrastinate is supported by experience that some of the future disasters we prepare for do not occur. All this makes a conservation policy difficult to introduce, and it is in part perfectly rational behavior by the public which makes it so.

Conservation has been roundly endorsed in part because the United States is said to be "wasteful" of energy. The alleged wastefulness is important. It implies that energy consumption can quickly and easily be reduced. It also gives the call for conservation a moral underpinning, since who can be in favor of waste?

Thus it is important to examine the factual basis for the alleged waste. Of course, we all observe some waste around us, and this evidence of the senses ought not be wholly discounted. Certainly there is waste, but the question is, Is there lots of it? Some of the evidence for the proposition that Americans are great wasters of energy is not persuasive for several reasons. First, the United States economy has been the most productive in the world for centuries, and this has not occurred through waste.

Second, the gross figures do not suggest major waste. The United States produces about one-third of the world's GNP and uses about one-third of the world's energy. Why are we said to be wasteful and everyone else not? In part it is because the United States is often compared with *selected parts* of the rest of the world, e.g., the United States with Sweden. But if this is a sensible procedure, why not compare Sweden with part of the United States, for example, Connecticut? Connecticut produced $18.40/MM Btu in 1971 while Sweden produced only $14.00 and the United States as a whole about $11.00. The difference between the United States and Sweden was 18 percent while the difference between Louisiana and Connecticut was 511 percent. Why not say both Louisiana and Sweden ought to be like Connecticut instead of saying that the United States ought to be like Sweden?

The criticism of such comparisons does not imply a ridicule of these studies. International comparisons are hard to make. For example, it is said that the United States could cut its energy consumption in half because the West German standard of living equals ours but uses half the energy. But closer examination suggests that about 40 percent of the difference in the household sector is due to Germany's smaller houses and greater population densities. Maybe these things do not affect your standard of living, but they do enter mine.

*If* Americans are not as wasteful of energy as we have been told, how much short-run possibility is there for reduced energy use without reduced output and standard of living?

Finally, how much difference can conservation make? As a quick example of the nature of the issue, some people estimate that there are about 2 trillion barrels of oil remaining to be produced. In 1975, 19.5 billion barrels were produced. Consumption at this rate would exhaust the estimated stock in 103 years. If worldwide energy *growth* continues at its historic rate of about 7 percent annually, the stock would be exhausted in about 30 years; if at 4 percent, in about 40 years. Thus, if exhaustion will be a *disaster*, an instant halving of growth rates—certainly more than can be achieved without major changes in current standards of living—will delay it only about 10 years. People might quite sensibly say the gain is not worth the cost in this case.

If we include the roughly 5 trillion barrels of "unconventional" oil (shale, tar sands, etc.), the answers are equally interesting.

$$7 \text{ trillion} \; = \; 359 \text{ at 1975 rates}$$
$$= \; 48 \text{ years at 7\% rate of growth}$$
$$= \; 70 \text{ years at 4\% rate of growth}$$

Thus we see that under certain conditions conservation can make a big difference. For example, if consumption could be reduced to, and maintained at, 1975 levels, estimated stocks could be stretched to 100 to 360 years instead of 30 to 70 years, *if unconventional oil can be produced.* Arithmetic like this is a partial explanation of why conservation is so appealing. This example also shows that conservation makes a bigger difference in years, the larger the remaining stock of oil. For this reason technology has a big leverage also, since technology enlarges the amount available to be conserved. Learning to recover oil from shale, tar sands, etc., will make a big difference. Of course, no one knows how much oil remains to be produced. Our history is one of *vast underestimates,* however. If we are vastly underestimating or overestimating remaining oil reserves, conservation may make little difference.

What are the conclusions from all this?

1. The lack of concensus on what conservation means and on why it is better to shift consumption forward to the future and the difficulty for an elected government to follow major policies whose results are slow in coming are all impediments to a major conservation effort.

2. Americans have not been shown to be grossly wasteful of energy. We do use a lot, to produce a lot, but it has not been shown that major amounts of consumption can be eliminated—beyond what is economically efficient—without significant impacts on output and standard of living.

3. As a result, while conservation should be pursued, policies to enhance production must complement it. Proposed energy policy to date has not been sufficiently oriented toward enhancing production and technology, but this may be changing.

# 7 Refuse Power and Its Immediate Impact on Energy Conservation

*K.S. Feindler*

When considering energy choices for the future, we usually hear about nuclear power, coal, and lately even solar energy. Refuse power is hardly ever discussed because, according to some technicians and politicians, it supposedly amounts to "only peanuts." It has even been overlooked as a topic to be considered by this conference.

Let's set the record straight. Refuse power can make an immediate and significant contribution to both our economic crisis and our energy crisis, which, by the way, are really interrelated.

Refuse power means the direct conversion of our urban and agricultural wastes into useful forms of energy in the form of steam, hot water, and electricity. It is best accomplished by placing these wastes in their raw form into special grate-boiler combinations, where under controlled conditions they are carefully combusted to produce steam for conventional power uses.

This technology is not new. It is well proved. Approximately 250 refuse power plants operate today throughout Europe and Asia. Considering that the oldest plants in Denmark have been in continuous operation for 40 years, no additional research and development is required. Often, these European refuse power plants are located within the population centers. In this way, they reduce energy consumed by transportation and, most importantly, permit cogeneration where electricity and district heat are produced in one and the same plant, thus resulting in an immediate energy savings of 30 to 35 percent. Cogeneration, unfortunately, does not appear as a topic of today's discussions either. European refuse power technology has been licensed to three major and well-known United States firms for immediate and nationwide application.

The impact of refuse power can be best illustrated by referring to New York City. With 20,000 short tons per day (STPD), New York City undoubtedly qualifies as the garbage capital of the world. Its population of 7,892,000 generates about 5 pounds of waste per capita-day. The heating value of this waste is such that one ton of waste equals one barrel of fuel oil.

A typical European refuse power plant can process 1600 STPD of mixed municipal solid waste into 40 megawatts of electricity. It takes three years, 1.2 million man-hours and $50 million to build such a plant. It has a minimum life span of 25 years which exceeds that of most gas and oil wells.

Considering these facts, one can readily estimate the potential impact of refuse power on New York City if the European approaches are to be followed:

New York City needs about twelve such plants.

Refuse power could help New York City reduce its oil imports by 7.3 million barrels per year.

Refuse power, in combination with cogeneration, can raise this potential to 10 million barrels per year.

Significant monies will be withheld from OPEC to reduce our balance-of-payments deficit:
$95 million (with just refuse power)
$130 million (with just refuse power plus cogeneration)

Some 1200 permanent jobs would be created for people directly involved with energy production in New York City.

Without considering engineering, manufacturing, and administrative efforts, 1000 new construction jobs could be created for the next three years.

Such a massive construction program would require about $200 million per year for 3 years. It is not inexpensive, but then it is comparable to other environmental and energy efforts:
EPA in 1978: $ 5.0 billion mainly for pure water
DOE in 1978: $10.6 billion mainly for nuclear and coal

Significant other benefits such as taxes could be accrued to New York City.

From the New York City example, you can extrapolate the impact of refuse power on our entire nation. There is a mountain of 300 million tons of garbage alone to be converted into energy. The United States EPA and the Batelle Institute have estimated that upwards of 250 such similar projects appear feasible in the United States during the next ten years.

Europe, of course, is not standing still. The latest trend is to combine refuse power with codisposal; i.e., refuse as a primary fuel is used to convert sewage sludge as a secondary fuel. This bottle which I carry in my pocket contains such sludge fuel with a heating value of 10,000 to 12,000 Btu per pound, which is comparable to that of coal. It came out of the Krefeld plant which serves 350,000 people on the west bank of the Rhine River.

Why haven't we succeeded so far? Simply because our attitudes as individuals are not attuned to systems thinking, and we lack the nuts-and-bolts approach to energy conservation. Simply because our institutional policies prevent rapid integration of our environmental energy and economy planners.

# Part III
# Coal as a Source of
# Energy

# Expanding the Use of Coal

## Arthur M. Squires

There was an "energy crisis" in England during the 1600s. Wood had become scarce and expensive, and techniques and devices were not yet in place for wide use of coal. As Maurice Adelman likes to point out, one of the great advantages of migrating to America was that you were going to a land where wood was plentiful. Our current energy crisis in a nutshell: For the first time since the 1600s in England, the cost of energy at the margin of the supply is now greater than the average cost of energy.

For the first time since the mid-1600s in England, our energy supply is vulnerable to shortages of a particular fuel required by devices not capable of switching to another fuel.

The consumption of coal in the United States has been static since about 1910. This has been the century of oil and gas. How can we use coal to replace these clean, convenient fuels? How can we again drive economic growth with coal? How do we expand the use of coal?

"Big technological fixes" will not restore a cheap energy margin. No doubt a great deal more coal will be converted to electricity by the year 2000, and it should be. No doubt some coal will be gasified, but more to supply low-Btu or intermediate-Btu gas to industry than to supply synthetic natural gas. Not much coal will be liquefied.

This is not, of course, to say that R&D on gasification or liquefaction should be diminished. Choices made in these areas by the year 2000 will affect energy supplies through the first half, at least, of the next century. The flash hydrogenation work at The City College Clean Fuels Institute is one of the most significant leads toward both cheaper gas and cheaper liquids from coal. Figure 8-1 gives data illustrating the sort of yields that are available from heating coal almost instantaneously in hydrogen gas at 100 atmospheres to the temperatures shown, and then quenching the temperature of the products after only a few seconds. Between about 700 and 800°C, there are very interesting yields of light aromatic hydrocarbons, suitable for chemical feedstocks or upgrading by relatively cheap procedures to gasoline. The liquid yield is better for some coals than for others, and Bob Graff and his team at The City College Clean Fuels Institute (CCCFI) have observed yields as high as 35 percent from some coals. The other products are a very good SNG and a coke, sufficiently low in sulfur from some coals as to be burnable with no controls on sulfur emissions. A number of companies and research laboratories have followed the CCCFI lead,

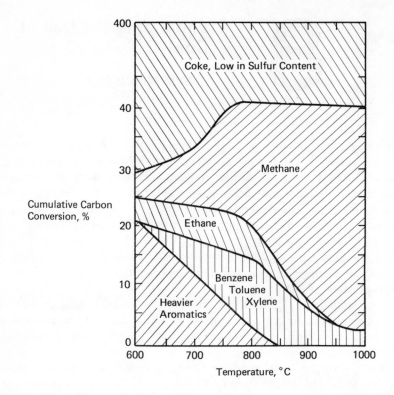

Note: Potential yields of BTX from some coals run as high as 20 percent, and potential yields of total liquid run as high as 35 percent.

**Figure 8-1.** Yields of BTX (benzene, toluene, and xylene) from Flash Hydrogenation Experiment at The City College of New York.

which began in 1971, in flash hydrogenation: Oak Ridge and Brookhaven National Laboratories, Rocketdyne, IGT, Cities Service, the National Coal Board of Great Britain, and (according to the grapevine) at least three or four other private companies.

The catch in all this is, How do you heat the coal and quench the products quickly enough? The problem is to feed a tremendous quantity of coal and hydrogen gas into a very small space, get the products back out again in a few seconds, and keep straight and proper what is going on in the small space. This is a chemical engineering challenge for which there is not yet an obvious winner. The answer almost surely lies in use of a fluidized bed at high gas velocity. The City College Clean Fuels Institute's test stand, in the hands of Joe Yerushalmi and his people, for study of "fast fluidization" and other high-velocity fluidization regimes will surely contribute to the answer.

There is a cheap, clean energy margin at Blacksburg, Virginia. It is located next to the Jefferson National Forest, and the cheapest energy in Blacksburg is wood. Of late people who own all-electric homes have been buying wood stoves like hotcakes. Dealers couldn't keep stoves in stock last winter.

The next cheapest energy in Blacksburg is coal. The Virginia Tech power station burns low-sulfur coal at $1.16 per million Btu's, about one-half the cost of fuel oil. This is in a chain-grate stoker that is environmentally acceptable.

With low-sulfur coal available to the Southeastern United States at this price, delivered in industrial-size lots, why aren't more industries burning coal for space heat and other relatively small energy needs? If people will buy wood stoves, why doesn't someone organize distribution of coal for the little guy who wishes to heat an apartment house or a home?

In short, there is a cheap energy margin in the Southeast which is going begging, to wit, the combustion of coal on a relatively small scale, say, from about 0.3 megawatt thermal to 75 megawatts (i.e., from 1 million to 250 million Btu's per hour), and perhaps including the 5- to 50-kilowatt range appropriate to a home or apartment structure.

This resource represents a relatively small part of the energy supply to only a quarter of the nation, but nonetheless a vigorous effort to put coal into this supply could have a catalytic effect, leading to rapid increase in coal elsewhere, and on a scale large as well as small.

It is easy to list physical factors that are missing: supplies of coal of the right kinds, a network of coal distributors and retailers, manufacturers of coal furnaces and dealers—all of which must be in place before large numbers of willing customers can be served.

Consider the problems of attitude that make it so hard for us to imagine rapid expansion of coal.

As a nation, we remember the dirt and labor of most small coal furnaces of the past. We remember the smoke and soot of cities like St. Louis and Pittsburgh.

As a nation, we forgot how to burn coal except on the largest utility and industrial scales. Only in boiler and electric companies are there corporate executives who know much about burning coal. Knowledge of coal in industrial firms, when it exists at all, tends to remain at the power-plant-engineer level.

Very few people knew then, or remember now, that the hard-pressed coal industry brought clean-burning, smoke-free, nearly labor-free underfeed stokers onto the market for household heat in the late 1940s. Only a few copies were sold before gas and oil knocked out coal from all except the biggest applications.

Almost no one in the United States knows that the British coal industry, pressed by Britain's Clean Air Act of 1956, followed the United States industry's lead in developing clean-burning "smoke-eaters" to make bituminous coal acceptable for home heat. The new furnaces have now achieved a significant penetration of the British market for home heat, displacing the more expensive clean-burning fuels, anthracites and coke.

Even to catch up with what we once knew about small furnaces, we badly need opportunities for "show and tell." Talk alone will convince no one that coal can be clean. This is especially true of the new fluidized-bed combustion practices. A confusing element is that there are a number of these practices, each with advantages and disadvantages, and there is risk that improper selections will be made before the differences are generally understood.

Virginia Tech is establishing a Coal Combustion Workshop to bring together about ten to fifteen examples of state-of-the-art coal furnaces. We will use the workshop to offer short extension courses on clean coal combustion in front of working furnace examples. The furnaces would be engineered for practical use and would range in size from a household furnace as small as 5 kilowatts to an industrial furnace as large as 3 megawatts. The workshop's primary function would be teaching, not research.

First, what is fluidization? In Germany in 1922, Winkler invented a new technique for gasifying coal. He saw it into commercial use by about 1927. His idea was to increase the rate of flow of gas upward through a bed of coal particles to the point at which each particle in the bed was buoyed by the rising gas, and then to a gas flow well beyond that point. When the pull of gravity downward on each particle was just canceled by the upward drag of the current of gas, the particles became free of one another, and the mass of particles took on much of the character of a liquid, finding its own level and possessing hydrostatic head. As the gas velocity was increased further, much of the gas rose through the "fluidized" mass of coal particles in form of bubbles, and the bed took on the character of a vigorously boiling liquid.

Figure 8-2 shows an application of Winkler's idea to coal combustion. This is the 1.4-megawatt fluidized-bed boiler operating at Pope, Evans and Robbins' laboratory at Alexandria, Virginia, in 1966. (We owe much to the late John Bishop, who put much of himself into this unit and is one of the heroes of the fluidized-bed combustion development.)

The fluidized bed in figure 8-2 consists almost entirely of a noncombustible solid. There are just a few particles of coal floating around in the bubbling, "boiling" mass of particles that you can think of as being sand for now. (Indeed, the British, who began the work on boilers of this type in the early 1960s, used sand. One of the heroes of the British effort was Douglas Elliott.) The sand serves two major functions. It provides a reservoir of heat which almost instantaneously ignites a coal particle added to the bed. The sand also carries the heat away from the narrow region of intense heat, just at the burning coal particle, and conveys the heat to the surface of metal tubes containing water to receive the heat.

Wayne McCurdy of the Department of Energy and the firm Pope, Evans and Robbins deserve enormous credit for having picked up the British lead so quickly—with the original target of developing a package industrial boiler shippable by rail. John Bishop and Ernie Robison of Pope, Evans and Robbins

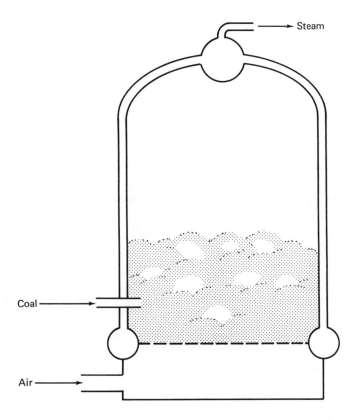

Note: The coal-firing rate is about 1.4 megawatts (thermal)—about the right size for a typical chicken farmer to heat his brooder houses.

**Figure 8-2.** Pope, Evans, and Robbins Experimental Rib at Alexandria, Virginia, in Operation in 1966.

quickly proposed imposing yet a third function upon the "sand" of the fluidized bed. They conducted experiments using limestone in the bed of figure 8-2 instead of sand, and they showed that the stone has the power to absorb sulfur dioxide generated during the combustion.

In the context of figure 8-2 and its initial research objective, the interest in small-scale coal combustion was prompted as follows. It was a very near thing last February for scores of millions—if not indeed hundreds of millions—of chickens. The chicken breeder industry—75 percent of it—depends exclusively upon propane for fuel; 75 percent of this industry has no alternative as the breeder houses are now built. A baby chick, in its first several weeks, *must* have heat. Propane inventories got very low in the East last February, and if a warming trend had not appeared near the middle of that month, personnel at the College of Agriculture at Virginia Tech warned of a lot of dead chickens. Our nation's consumers might have had a rude shock at the price of chicken meat if the cold had lasted a few weeks longer.

The chicken farmer typically needs heat at a scale of about 0.3 to 2 megawatts thermal. The 1966 unit seen here, for 1.4 megawatts, would do nicely for many chicken farmers.

The British have given more thought to equipment for small-scale use. Figure 8-3 is a vertical fire-tube boiler for 3.5 megawatts developed by Raymond Hoy and his team at the National Coal Board's laboratory at Leatherhead, south of London. There are three rows of horizontal tubes removing heat directly from the 43-inch-wide bed, and vertical tubes in a tank of water that further recover heat from the combustion gases. A striking feature is that the unit can burn oil or gas as well as coal. This unit is licensed for manufacture in the United States by the Johnson Boiler Co. of Michigan.

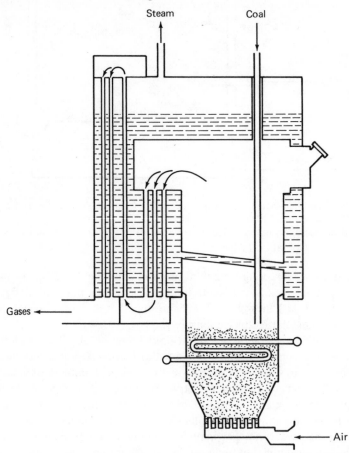

**Figure 8-3.** Vertical Fire-Tube Boiler for about 3.5 megawatts (thermal) Developed in the National Coal Board's Laboratory at Leatherhead, England.

Figure 8-4 shows the pattern of solid circulation that the fluidization engineer ordinarily expects to see in a deep fluidized bed. Bubbles in the gas flow tend to move to the center, creating a geyser up the middle, with a lazy downward return of solid at the walls.

Figure 8-5 is a furnace developed by the late Douglas Elliott's team at Stone-Platt Fluidfire near Birmingham, England. There is a reverse pattern of solid circulation, with a strong downward current at the middle and a lazy upflow in the wings. This reverse pattern leads to long residence time for fuel particles in the bed, and the furnace is particularly effective in burning fuels which produce fine, light particles that are difficult to burn. It has been tested successfully on paper wastes, fabric wastes, chopped rubber tires, wood and wood wastes, as well as coal.

This design was selected for the first purchase for Virginia Tech's coal combustion workshop. The furnace is about to undergo start-up trials at the manufacturer's shop in England, and it should be shipped in a few weeks. Its rating is 0.3 megawatt.

Figure 8-6 shows another novelty, discovered and exploited by John Highley and his team at the National Coal Board's laboratory at Stoke Orchard near Cheltenham. Lumps of coal will float and burn nicely upon a shallow fluidized bed of sand.

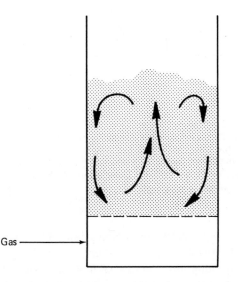

Gas ⟶

Note: Bubbles in the gas flow tend to move toward the center of the bed, creating a strong upward motion there, with a lazy downward movement of solid elsewhere. The "geyser" in the middle tends to hurl fine particles of fuel out of the bed, reducing combustion efficiency.

**Figure 8-4.** The Pattern of Solid Circulation Normally Expected in a Deep Fluidized Bed.

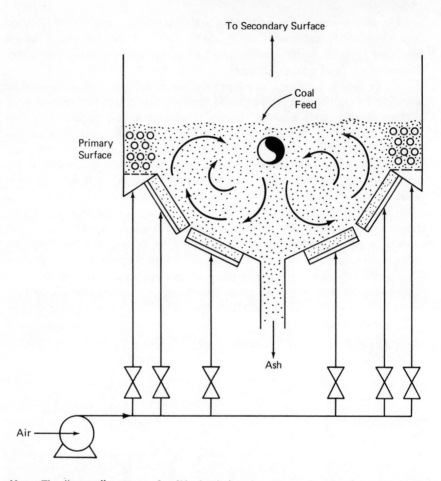

Note: The "reverse" pattern of solid circulation—i.e., strong downward current at the middle and lazy upflow elsewhere—is conducive to long time of residence of fine, light particles of fuel in the bed. The furnace has been tested successfully on paper wastes, fabric wastes, chopped rubber tires, wood and wood wastes, as well as coal.

**Figure 8-5.** Fluidized-Bed Furnace Developed by Stone-Platt Fluidfire Ltd. of Dudley, West Midlands, England.

This is a strong candidate for burning low-sulfur stoker coal at capacities where the underfeed stoker might previously have been selected. With respect to the problem of burning bituminous coal in a fixed bed, when you put a fresh lump of coal onto a grate furnace, tar matter distills and cracks and repolymerizes to form soot and smoke. Power engineers all the way back to James Watt understood perfectly well how to get rid of smoke, and at the same time to utilize the heat value present in smoke, which represents a serious thermal loss if

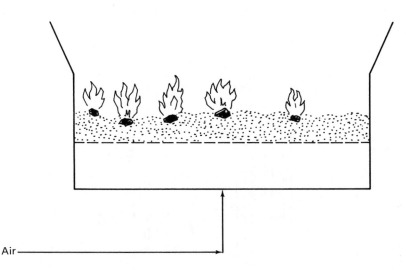

Note: The volatile matter in a fresh lump of coal distills over several minutes. The lump moves in a random motion across the surface of the bed during this time, while the volatile matter burns brightly in a flame just above the bed. There is little splash of solids from the surface of the bed, and it can be installed in a space of relatively small height.

**Figure 8-6.** Lumps of Coal, Floating and Burning upon a Shallow Fluidized Bed of Sand.

it is allowed to go up a chimney. What you do is provide a zone of intense secondary combustion, beyond the fuel bed, to burn up the smoke. People burning bituminous coal in simple small furnaces didn't do this; hence, smoky Pittsburgh, St. Louis, London, etc. The underfeed stoker provides such a zone of secondary combustion. So do the National Coal Board's new "smoke-eaters." Good small-scale devices have existed for some time, making it possible to burn bituminous coal without smoke. The problem in the bad old days of coal was that they were not used.

Figure 8-6 shows another way to burn without smoke. A lump of coal, 1 inch or so in size, takes a number of minutes in distilling its tar. It floats around over the surface of the fluidized sand, and the tar vapor burns brightly in a flame just above the sand surface.

One of the main advantages of Highley's shallow bed is that there is little splash, and the bed can be installed in a small height. Figure 8-7 shows an application, where the shallow bed has been used to retrofit a horizontal fire-tube boiler, previously firing oil or gas, for coal firing. Figure 8-8 is a cross-sectional view of the horizontal fire-tube boiler fitted with a shallow fluidized bed.

Figure 8-9 is a shallow bed with a large excess flow of air. It furnishes hot

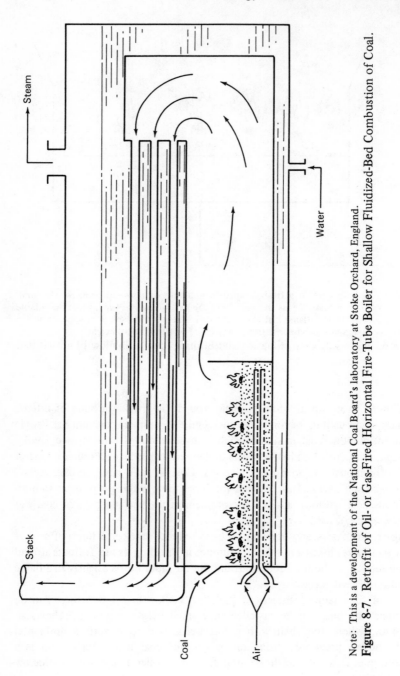

Note:   This is a development of the National Coal Board's laboratory at Stoke Orchard, England.

**Figure 8-7.**   Retrofit of Oil- or Gas-Fired Horizontal Fire-Tube Boiler for Shallow Fluidized-Bed Combustion of Coal.

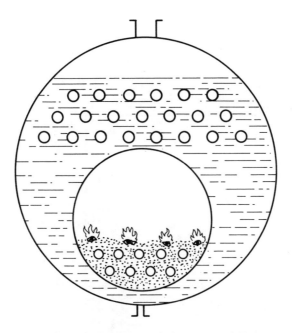

**Figure 8-8.** Cross-Section through the Boiler of Figure 8-7.

gases to a rotary drier for dehydrating agricultural products such as alfalfa and grass. There is a 5-megawatt unit of this design near Nottingham, England, whose chronology is as follows. Highley began work on the shallow-bed combustion idea in 1972. The Nottingham unit was the third put in; it was ordered in January, started up in mid-April, and by late June the customer was delighted, with both its reliability and its drying costs.

This chronology, and the account of how Winkler brought fluidization to commercial use at the 100-megawatt scale in just five years, illustrates that you can develop and commercialize on a small scale far quicker than on the larger scales with which we have been more familiar in recent government R&D programs.

Figure 8-10 is a "smoke-eater" from South Africa, very different from the National Coal Board's designs, and its novelty suggests that there may be a great many other novel and useful configurations yet to be invented. The spring urges the bed of coal toward the right. Air enters the fuel bed by natural draft through grate bars at the lower right, and the zone of combustion is generally to the right of the sloping dashed line. It is just about at this line that a lump of coal heats up and releases its tar vapors by distillation. These vapors must pass upward through the bed of burning coals and into a narrow passage at C, which provides a zone of turbulent combustion to destroy the tars. The narrow passage traverses a block of refractory. Hot combustion products from the narrow passage flow

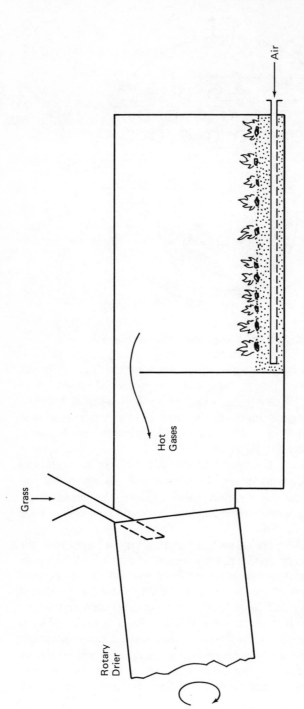

Note: This is another development from Stoke Orchard. Three commercial units are in operation in England, the largest at 5 megawatts (thermal).

**Figure 8-9.** Shallow Fluidized-Bed Combustion of Coal for Dehydrating Alfalfa and Grass.

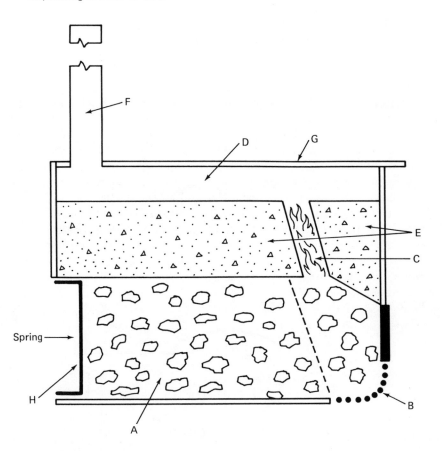

A = coal supply. The dashed line generally indicates zone where volatile matter distills from the coal.

B = grate bars for introduction of combustion air by natural draft. Primary combustion of coke occurs in zone at right of dashed line.

C = turbulent zone for cumbustion of volatile matter to produce smoke-free combustion products.

D = space for cooling combustion products by transfer of heat to G.

E = block of refractory with duct C.

F = flue.

G = metal plate for cooking food.

H = ram to push coal from supply A toward primary combustion above B. The ram might be driven, for example, by a spring (not shown in drawing).

**Figure 8-10.** Smokeless Kitchen Stove Developed by Fuel Research Institute of South Africa at Pretoria.

beneath a metal plate G, on which one can cook food. The Fuel Research Institute of South Africa at Pretoria developed this kitchen stove for the black townships, which are today enveloped in smoke from South Africa's subbituminous coal.

The chicken farmer does not use a large number of quads of energy. He burns propane today equivalent to about 750,000 tons per year of coal. Conversion to coal would not greatly increase coal use, but might nevertheless be important for ensuring the robustness of an important food supply.

The last furnace example leads to a potential use for coal in small-scale combustion on a scale to rack up a significant number of quads.

Figure 8-11 is a 25-megawatt furnace just now being commissioned by the District Heating Scheme of Enköping, Sweden. This most flexible unit is intended to burn high-sulfur residual oil (using limestone to capture sulfur), coal, peat, wood and wood wastes, and municipal wastes. There will be a bag filter to capture dust. The design was developed by a Norwegian firm, Mustad Støperi of Gjøvik. This firm hired a team of college professors to lead the development, at Trondheim Technical High School. The work began in 1969 and led in May 1975 to the first commercial unit, at 2 megawatts. The unit has two unusual features. It provides for combustion in two stages—primary combustion in the fluidized bed and secondary combustion in an unusually tall space above the bed. It recycles a part of the combustion products to the primary air, returning part of the heat to the bed. These features lead to negligible emissions of

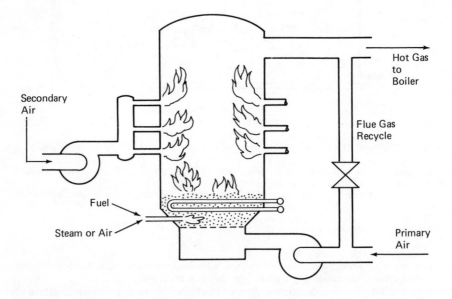

Note: The practice is characterized by secondary air overfiring and flue gas recycle. An example at 25 megawatts (thermal) is just now being commissioned by the District Heating Scheme of Enköping, Sweden.

**Figure 8-11.** Fluidized-Bed Combustion, Practice Developed by Mustad Støperi & Mek. Verksted A/S of Gjøvik, Norway, in Collaboration with Trondheim Technical High School.

nitrogen oxides and to capability to burn fuels of low and varying heating values, such as wet industrial or municipal wastes. It is a flexible unit, and it is an impressive achievement by a small country that was a late starter in the fluidized-bed-combustion development race.

District heating is a major opportunity for expanding use of coal. Figure 8-12 shows the hot water pipes of Enköping, Sweden, installed since 1970, and now furnishing more than 75 percent of the space-heat requirements of this small city. It measures about 1.5 by 2.5 miles and has 20,000 people. The current load is 75 megawatts. The map illustrates something about how a district heating scheme like this develops and grows. Swedish experts emphasize that the scheme *must* start small. The two outlying fragments illustrate how a scheme might begin, with several small units—from 2 to 5 megawatts, say—and with the intention of trading in these units after the scheme grows. There are 25- and 35-megawatt hot-water furnaces at the other two Maltese crosses, which will soon be gone, since the open Maltese cross marks the site of a new, permanent 100-megawatt heating plant, next to a canal for delivery of coal in barges.

This technology does not involve cogeneration of heat and electricity, but simple hot-water furnaces with no other function. The incentive to put in such furnaces today in Sweden is simply that they can make heat cheaper. This is largely a matter of thermal efficiency. The household furnace is seldom better than 50 percent efficient—almost never better than 60 percent. Enköping's scheme today runs around 82 to 83 percent. There is a stack loss of about 10 or 11 percent, and about 7 percent loss of heat in transmission. The national average efficiency of *all* district heating schemes in Sweden, including some rather poor ones, is 80 percent. That is heat paid for by customers, based upon heat meters, divided by heating value of fuel burned.

It should be emphasized that district heating schemes must start small. It is next to impossible to line up all the customers in a large district all at once, signing them up for district heat, even if you could miraculously lay all the necessary pipe in an instant.

After a scheme is established, it does become an exciting heat sink for electricity generation. Figure 8-13 illustrates how Swedish heat engineers, in their long-range thinking, contemplate cogeneration equipment whose sole function is heating, with electricity going to rural districts and hot water to cities.

Figure 8-14 is an overlay of Enköping, Sweden, on Blacksburg, Virginia, also a city of 20,000, but with another 20,000 students—many living in townhouses and apartments of recent construction that are candidates for retrofit to district heat. Perhaps the earliest, small furnaces should be oil-fired, but in the long run, as the system grows, it provides an alternative scenario for converting heat from oil and gas to coal.

How does this scenario compare with synthetic gas and with electricity? For starters, the efficiency is roughly twice as good. For finishers, the capital cost

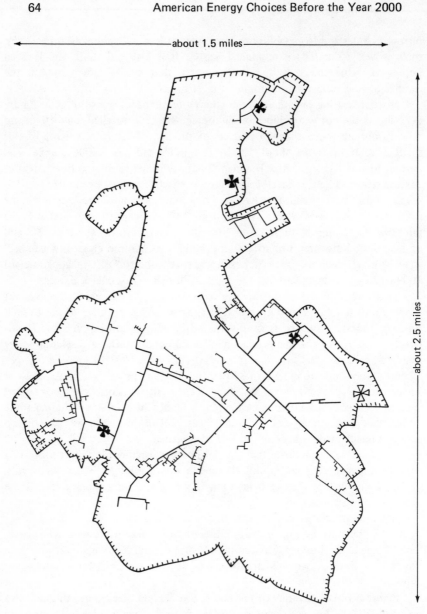

←——————about 1.5 miles——————→

about 2.5 miles

Note: This is a town of 20,000. The closed Maltese crosses are existing oil-fired water-heating furnaces. The open Maltese cross is the site of a new 100-megawatt station just now being commissioned. The other furnaces will be traded on a second-hand furnace market.

**Figure 8-12.** Schematic Illustration of District Heating Piping Network at Enköping, Sweden, and Approximate Limits of Urban Development.

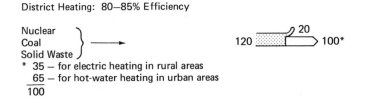

District Heating: 80–85% Efficiency

Nuclear ⎫
Coal    ⎬ ——————▶                    120 ▒▒▒▒▒▒ ⟹ 100*
Solid Waste ⎭                                    20
* 35 — for electric heating in rural areas
  65 — for hot-water heating in urban areas
 ———
 100

*The asterisk is to remind us that if even a relatively small part of the electricity is used in heat pumps, the 120 units of fuel energy can drop below 100. That is to say, the apparent efficiency can be greater than 100 percent.

**Figure 8-13.** How 120 Units of Fuel Energy Can Be Converted to 100 Units of Space Heat.

appears to be less than half. The piping in Enköping could be replaced at today's prices for about $70 per kilowatt (thermal). A unit for coal combustion at the 100 megawatts (thermal) that Enköping will need would cost about $100 per kilowatt including all environmental controls and coal receiving, handling, and stockpiling.

There may be nothing we could do that would be more cost-effective on behalf of our city environment, or that we could do more quickly, than converting space heat from small oil fires to district heat. Consider Professor Oden's famous maps of the acid rainfall of Northwest Europe. Figure 8-15 for 1956 shows a region of rain with pH less than 5. In figure 8-16, for 1959, this has grown a bit, and a region of pH less than 4.5 has appeared. In figure 8-17, for 1961, we have a region of pH less than 4. And in figure 8-18, for 1966, the region less than 5 and the region less than 4.5 have grown enormously, while that less than 4 persists.

But these were just precisely the years during which Northwest Europe closed down its small-scale coal combustion in commerce and industry, and moved to oil. (See table 8-1.) Total use of coal declined, as large electricity plants became the major purchasers of coal, and oil increased more than fourfold, exceeding coal in 1966.

And so, you ask yourself, What about production of sulfur trioxide from oil and coal combustion? Then you remember that the acid dewpoint of a small oil boiler is about 30°C higher than that of a small coal-fired boiler. Then you look in the literature, and although this is limited in amount, it seems pretty clear that a small oil furnace is a fairly dirty device—dirty, that is to say, for its emissions of sulfur trioxide in form of a sulfuric acid mist.

What is even more striking is that the sulfur trioxide emissions from a small oil furnace do not appear to track the sulfur content of the fuel. (See figure 8-19.)

**Figure 8-14.** Comparison of Extent of Urban Development at Enköping (hatched curve) with Blacksburg, Virginia, a Community of 20,000 Residents and 20,000 Students.

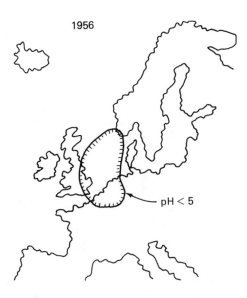

**Figure 8-15**. Odén's Data Showing Extent of Acid Rainfall in Northwest Europe in 1956.

It should be pointed out that large utility boilers can easily be operated, at low excess air, to emit only about 1 or 2 parts per million of sulfur trioxide. A typical coal-fired utility boiler might emit 5 parts per million. The data in figure 8-19 were obtained under laboratory conditions in a clean, well-maintained furnace. There are reasons to suspect that a typical New York City apartment house furnace might emit a great deal more—even when burning oil at 0.3 percent sulfur. We do not know. We do not have data.

The Environmental Protection Agency might well consider a test program on sulfuric acid emissions from apartment house furnaces in New York City if it wishes to understand New York's atmosphere and the health effects that it causes. New York burned oil at about 2.5 percent sulfur in 1966 and oil at 0.3 percent sulfur beginning in 1971. We brought down the sulfur dioxide levels in New York in this proportion. But it would appear that soluble sulfate particulate matter did not come down by much. Acid rain is an area problem, here as in Northwest Europe. So, it would appear, is the problem of soluble sulfate particulate matter. But small oil fires are pretty ubiquitous. Have we gone after the wrong villain? We do not know. We wish we knew. We wish that the EPA would spend just a fraction of the research dollars it is investing in its conclusion

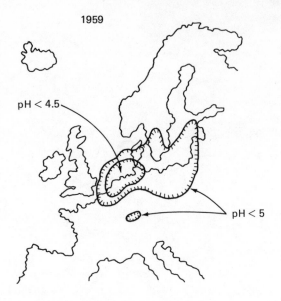

**Figure 8-16.** Acid Rainfall in 1959.

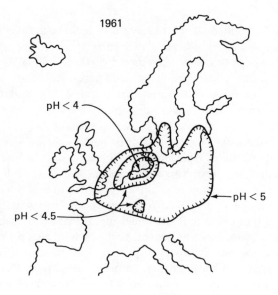

**Figure 8-17.** Acid Rainfall in 1961.

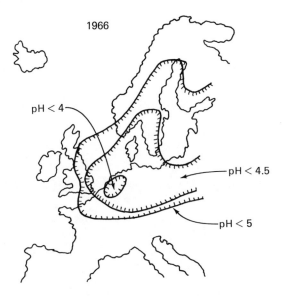

**Figure 8-18.** Acid rainfall in 1966.

that New York City's problem arises from the tall stacks of coal-fired power stations along the Ohio River into the hypothesis that at least a significant part of New York City's problem is home-grown. We should emphasize the word *hypothesis.* We just do not know. No one does.

We should weep over the latest Clean Air Act Amendment and its imposition of "best available control technology." This, of course, is a cop-out, and releases the EPA legal staff from any obligation to defend decisions on environmental control strategies. There is no longer any basis for a dialog on

**Table 8-1**
**Western European Fuel Consumption**
*(metric tons of coal or coal equivalent)*

|        | Solid Fuels | Liquid Fuels |
|--------|-------------|--------------|
| [1956] | [550]       | [150]        |
| 1958   | 540         | 189          |
| 1959   | 515         | 213          |
| 1961   | 532         | 270          |
| 1966   | 486         | 516          |

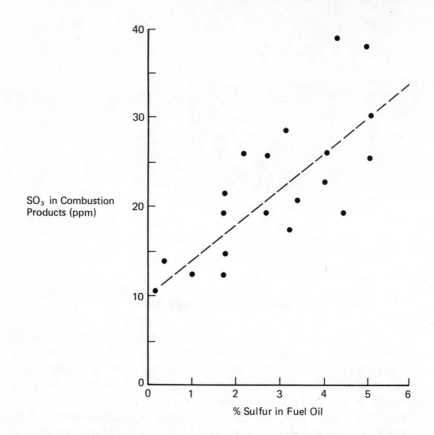

Note: There are reasons to suspect that emissions from a real-world furnace might be greater. (It is important to note that emissions of sulfur trioxide from a large oil-fired utility boiler can be maintained at much lower levels, around 1 to 2 ppm, by firing with low excess air, an impractical procedure for the operator of a small furnace.)

**Figure 8-19.** Emissions of Sulfur Trioxide (in Form of Sulfuric Acid Mist) from Careful Laboratory Tests on a Small Oil-Fired Laboratory Furnace.

costs and benefits of control strategies. The villains are whoever the EPA says they are.

Poor coal! We remember the sins of our fathers and grandfathers, who burned it with copious yields of smoke. The question should be raised: Do we also blame a lot on coal today that we really should lay at the door of oil?

# 9

# Coal Production and Protection of the Environment

*Carl Bagge*

It is an indication of the progress we have made in this country that we are able to mention coal and environmental protection in the same breath. Twenty or fifteen or even ten years ago, just the idea of a coal industry spokesman seriously discussing protection of the environment would have been dismissed as ludicrous.

But in recent years more and more Americans have come to the realization that greater coal use and a pleasant environment are not two necessarily incompatible extremes. It is particularly important that this realization grow at this time, because we have many difficult decisions to make over the next few years regarding coal, energy, and the environment.

The impression should not be conveyed, however, that the coal industry and environmentalists have courted, gotten married, and are happily living in suburbia. While the fact that a spokesman for a coal association is participating on a panel about the environment is a sign of progress, it is also an indication that there are still some serious problems—issues viewed with equal alarm and conviction on both sides.

It would be presumptuous to try to interpret the positions of environmentalists. They are more than willing to do that themselves. But it should be asserted that the coal industry, contrary to what some people would have you believe, recognizes that there are legitimate environmental concerns that have to be dealt with if we are going to use more coal in this country. There are, on the other hand, other environmental issues, supported with equal ardor by their creators, that are not so valid in our eyes and that are being carried too far. These are environmental issues of dubious value that could have serious, detrimental economic and energy consequences for this country.

The difference between these two types of environmental concerns is as wide as the gap of relative importance between animals traditionally associated with the endangered-species list, and those more recent additions. In the case of the former, we have such noble creatures as the bald eagle, buffalo, and whooping crane. In the case of the latter, we have the snail darter, which is a 3-inch-long minnow, and the furbish lousewort, a ragtag wild snapdragon—both rare and apparently confined to single locations in this country. But both have been the cause for recently stopping two separate federal hydroelectric projects.

This is not to say that this second group is not important. To be sure, to an environmentalist they are. But when relatively obscure causes begin to block energy projects of major proportion—whether they be coal or not—then we have a situation that is no longer comical. We then have a situation that is both serious and symptomatic of a major problem facing this country today. That problem is: How do we balance the energy needs of this country with reasonable protection of the environment?

Debate over this question has heightened somewhat since President Carter announced his national energy plan in the spring of 1977, stressing a conversion to coal as this country's main energy hope for the near future.

Obviously, the coal industry was very gratified that President Carter made both this commitment to coal and an attempt to establish some national energy priorities. Whether we all agree with the methods of attack chosen by the administration to deal with America's energy problems, we can certainly all agree that it is time to act. It makes no sense to deplete our supplies of oil and gas, which comprise only 7 percent of our domestic energy reserves, by continuing to use them to produce three-quarters of our total energy. It makes no sense to continue policies that result in coal, which comprises 80 percent of our energy reserves, being used to produce less than 20 percent of our total energy.

President Carter set, as a major goal of his energy plan, a doubling of present coal production to more than 1 billion tons by 1985. While administration spokesmen, such as Secretary of the Interior Cecil D. Andrus and Secretary of Energy James Schlesinger, have expressed confidence that the coal industry can meet this goal, and do it in a manner compatible with environmental protection, a number of organizations have publicly taken issue with the production aspect of the energy plan. These organizations—including the Library of Congress, the General Accounting Office, the Office of Technology Assessment, and the Business Roundtable—have said that either the demand for or supply of coal, or both, *could not* reach the levels contemplated in the Carter plan.

The American coal industry has great confidence that we can double coal production by 1985. The coal industry and its biggest customers have made commitments to mine and burn more than enough coal to meet the administration's production goal. But the question is not whether we can mine it, but whether it can be burned.

The National Coal Association recently undertook surveys on the anticipated future supply and demand for United States coal. The studies found that the coal industry plans to open or expand at least 332 mines, which will produce an additional 594 million tons of coal by 1985. In addition, electric utilities plan 241 new coal-fired generating units, which will use 400 million tons of coal by 1985.

Even from a conservative standpoint, the coal industry visualizes a potential domestic demand for coal in excess of 1 billion tons by 1985, not to mention export tonnages. It is clear from the trends that are occurring—such things as the very strong movement away from the use of oil and gas and toward the use of more coal by utilities—that private industry is doing about all it can to make the switch to increased reliance on coal. Billions of dollars are being committed for the opening of new mines, increasing coal production, and buying new coal-fired boilers and other equipment. But it is now up to government—not industry—to make these plans feasible.

While the coal industry is optimistic about its own capabilities and the prospects for its market, we have become very disturbed with the trend of recent actions by the federal government. While at one moment we hear rhetoric from government endorsing greater coal use, we see actions taking place the next that would achieve the opposite effect. And many of the actions are the result of questionable and misguided efforts by the government to deal with the issue of energy and environmental needs.

For many years, coal has been a demand-constrained resource. Our markets traditionally fluctuated in relation to the demand for our product by our major customers. We could mine all they wanted and more. But now government action—or inaction—has replaced demand as the principal constraint on coal production. And when we assess the future, with recent government performance as an indicator, we can only react with alarm.

The inability of government to deal fairly with the energy and environment question surfaces most frequently in much of the legislation passed by Congress. For instance, one of the most important constraints to greater coal use in recent years has been the Clean Air Act. There is no question that the burning of coal creates potential environmental problems, and the coal industry has no quarrel with air-quality requirements that are necessary to protect public health. But we often face state regulations or interpretations of the Clean Air Act that are far tighter than necessary to meet even the standards proposed by the federal government to protect health. To further complicate the situation, just last year Congress passed additional amendments on clean air, creating the very real danger that this constraint on coal demand could become even tighter. The full impact of these amendments will not be known until regulations under the new law are published. But it is clear these new amendments could have a heavy and adverse impact on both the kind and the amount of coal used in the United States.

Although it dealt with another issue that is an important and valid concern, the Coal Mine Health and Safety Act of 1969 unfortunately has proved to be another major constraint on coal production. While this law had some solid merit, it also had aspects that sent productivity down and costs up and that were counterproductive to its original aims in certain instances. For instance, the additional manpower and nonproductive procedures required by the law, which often have done little to promote a safe working environment, made coal the only major industry in which the output per man-day has fallen precipitously since 1969.

The act has reduced fatalities, but not eliminated them. Like so many other federal laws, it does not really get to the root of the problem. Coal industry studies show that most of the remaining accidents in the mines are the result of a lack of training and education, poor work habits, or a lack of motivation. The conditions of the mine and its equipment, which are strictly regulated by the 1969 law, are not the prime cause. The truth of this situation has even been recognized by Robert Barrett, who directs the government's mine safety and health agency.

Last year Congress saw a need to pass amendments adding even more stringent mine safety regulations. In this instance, the coal industry is again faced with unnecessary increased regulation that not only will not improve the working environment substantially, but will perpetuate and possibly even accelerate the decline in productivity that has occurred in our industry over the past nine years.

There has been much debate over another piece of legislation reflecting the energy and environment question—the new federal Surface Mining and Reclamation Act. While the coal industry did not favor national surface mining legislation, we accept the fact that we now have a federal law and we intend to comply with it. But we see the major impact of this law as simply complicating a job that the industry is already doing well—the effective reclamation of surface-mined lands.

The new regulations recently developed by the Department of the Interior to implement this law could make it difficult to produce from some coal reserves, could reduce production and productivity in those mines that can produce, and may sharply increase the costs of production. Also, there is great potential for delays in production while producers and the government find their way through the lengthy process of applications, reviews, hearings, environmental assessments, and other steps required by the law. Indications are that there will also be lengthy court delays while new procedures are tested by those who oppose increased coal production.

To make this situation even more difficult, the Department of the Interior has, since 1971, had a de facto moratorium on leasing publicly owned coal lands. It appears that a recent court decision in Washington regarding a suit brought by environmentalist groups may even further extend this moratorium. In the meantime, the failure to proceed with leasing is already beginning to affect some planned coal production in the Western United States, where most of this federal land is located. And the impact will grow as time passes.

It would be naive to say that these are the only major problems standing between the coal industry and achievement of President Carter's 1985 coal production goals. There are many others having nothing to do directly with environmental issues, including the availability of capital, transportation, and manpower. But we aren't afraid of tackling these problems, as long as additional restraints are not imposed from other sources. We are accustomed to meeting challenges and dealing with problems, and we are confident of our ability to handle those we can zero in on.

What we cannot handle, however, are government actions that do not consider the economic and other side effects of legislation in the same manner that they consider environmental impacts. It seems that over the past ten years, the government has supported one environmental cause after another without soundly examining the effects that might occur in other areas. And, as can be seen from the few examples cited here regarding the coal industry, we are paying a high price for what has been nothing more than environmental excessiveness in some cases. It appears the ante will continue to rise in the future.

Environmental values are important, and they have a significant place among our national concerns and priorities. But this nation will not achieve any of President Carter's major energy goals—including coal production—until its government leaders come to the realization that environmental issues can no longer be supreme. We will not be able to produce more of our energy from our own resources unless the government takes a more rational approach to environmental and energy issues. For this country to develop the resources it needs to meet a projected 4 percent annual growth in energy demand over the next decade, the environment can no longer be the overriding consideration in regard to energy projects.

With this in mind, the coal industry's biggest challenge over the next few years obviously isn't to mine enough coal to meet demand—we know we can do that. It is, instead, to convince government to broaden its approach when considering legislation and to strike a fair balance between environmental and energy needs.

The coal industry recognizes that it has an obligation to meet the nation's need for secured domestic energy. We intend to meet that responsibility, if legislative and regulatory requirements are reasonable. We also recognize that government has an obligation to allow us to proceed to do this job to the best of our ability. We cannot be at the same time the nation's energy warrior and its whipping boy.

As we enter a new year and the beginning of a new age for coal, we must continue to press for government—at all levels—to keep one eye fixed on the real world. We must press to ensure that just social objectives for clean air and water can be met without sacrificing the nation's need for secure energy supplies. We must press public officials to fully evaluate new legislative and regulatory proposals in view of their actual impact on the specific plans of industry. And we must urge government to weigh carefully the benefits of environmental protection against critical delays in energy development.

# 10 Capitalization and Financing of Coal-Fired Generating Plants

*Richard Disbrow*

The financing of a coal-fired generating plant today and in the years ahead is best discussed in the context of the broader question: Can coal-fired generation be provided on a timely basis to meet this nation's growing needs for electric power? Financing is one element in a multifaceted answer.

Ten years ago there would not have been the need for a seminar on American energy choices before the year 2000, and certainly the question of providing new generating capacity wouldn't have been of widespread interest—it was taken for granted. Today, energy and energy supply are a matter of continual and spirited discussion. Today, anyone concerned with the vitality of our economy must also be concerned with whether new coal-fired generation can be counted upon to assume a role of increasing importance in meeting our expanding needs for energy.

Much has changed in ten years. We have been through a period of rampant inflation and now look upon a 6 percent inflationary rate as the near-term floor. We have witnessed upheaval in the international energy markets in the aftermath of the oil embargo in the fall of 1973. We have seen the rise of a dedicated environmental movement which has placed great financial burdens on our industrial base and which, in view of its changing regulations, adds great uncertainty to our planning processes.

How does all this relate to new coal-fired generation? It can be illustrated best through example. For better than two decades after the close of World War II, electric utilities were the darlings of investor and consumer alike. Technological development and economies of scale led to decreasing unit costs of new facilities and periodic rate reductions. The common stocks of electric utilities sold at 20 to 25 times earnings. Electric utilities had enormous financial resources relative to their needs. Governors, legislatures, and the general public welcomed new utility investment. And the primary subject of today's meeting, coal, could be purchased for $4 per ton.

What of today? Economy of scale and technological development cannot compete with inflationary rates. The cost of new coal-fired generation as built in the mid-1960s has increased conservatively from $150 per kilowatt to $450 per kilowatt and is heading higher. To this must be added the cost of new environmental facilities which, in the case of scrubbers to remove sulfur dioxide in flue gases, may add 20 to 25 percent more. The common stocks of electric utilities sell at 8 to 9 times earnings. Electric utilities have very finite financial resources, siting legislation and environmental prohibitions make the location of new facilities difficult at best, and few remember they welcomed electric utility construction a few years ago. That $4 coal is now $20 to $25 per ton. And periodic rate increases, rather than rate reductions, are now required. There were few opponents of rate reductions. The same cannot be said for rate increases.

Let's assume that today we have made the decision to construct a new 1000-megawatt coal-fired generating unit and see how we will finance it and what impediments lie in the way of our financing program.

We'll assume we completed the two-year environmental impact statement and have received the multitude of required federal, state, and local permits with a minimum of intervention and delay. We are then a minimum of two years down the path from the point of decision and face at least a four-year construction program. Somewhere during that initial period we found it necessary to commit orders for major components in spite of uncertainty as to when or whether we could use them. But we did commit ourselves, and so we face large expenditures three to four years away and must develop a program to raise those funds on a timely basis.

For a 1000-megawatt plant costing $450 per kilowatt, we need $450 million compared to $150 million if we had built ten years ago. Three times as much! In the past we would have financed construction, on a temporary basis, allowing short-term debt to reach as much as 10 to 15 percent of capitalization before arranging permanent financing. Our financing program would contemplate the sale of first-mortgage bonds, preferred stock, and common stock to provide the external funds needed to finance the new unit. We would have had capitalization ratio targets of 50 to 60 percent long-term debt, 5 to 10 percent preferred stock, and the remainder, 30 to 35 percent, in common-stock equity. Except in unusual circumstances, retained earnings from current operations would suffice for the necessary common equity to support the new debt financing and maintain capitalization ratios. We would have our program in place and be underway, selling the securities as necessary.

Why isn't it that simple today? The reason is that if we are to maintain our target capital ratios, all else being equal, we need three times as much of everything to build our $450 million plant as to build the $150 million plant of ten years ago. Additionally, the debt interest rate is double what it was then.

The combination of amount and interest rate of debt produces a serious result. A general requirement in a utility's open-ended mortgage is that, as a prerequisite for issuing additional debt, earnings before federal income taxes be a minimum of two times long-term debt interest expense, including that of the issue to be sold. Our new plant requires three times as much debt at twice the interest cost and, therefore, six times the pretax earnings. In our 1000-megawatt example, this equates to $40 million incrementally to cover the additional interest charges. With nothing else changed, we now require $40 million more from our customers than previously. This seriously restricts our ability to issue new debt. In the past, pretax earnings of five to six times were common. With the change in financial parameters, we now have 2 to 2.5 times. This limits our financing flexibility.

Additionally, the decline in pretax earnings to debt interest has led to deratings of electric utility bonds from AA to A and, in some cases, to BBB. The lower-rated securities bear higher interest rates, which, when issued, use a larger portion of our remaining margins.

Therefore, we must reexamine our historic use of short-term debt. We can no longer accept the premise that we can temporarily fund construction with large amounts of short-term debt and then permanently fund the short-term debt. We may not have the means. We learned in 1974 the importance of liquidity in volatile financial markets. Short-term debt must be used sparingly and counted on primarily as the last, rather than the first, financing vehicle.

Some might suggest we avoid the coverage problem by changing our capital structure to increase the proportion of equity capital. But equity capital is more costly than long-term debt, and we now need to sell common equity frequently merely to maintain our historic equity ratio. Because of inflation of all costs, our earnings performance may not support even a modest increase in the equity ratio. To sell more than the minimum at or near book value, as the market now dictates, isn't feasible.

Perhaps we should depart from our standard financing practices and explore other financing avenues. Formation of a subsidiary to finance the construction out from under the existing mortgage might be considered. Project financing with an unrelated party doing the construction financing might be viable. At the completion of construction, we could buy the plant or, perhaps, even lease it. Superficially, these alternatives may seem attractive, but they merely avoid the basic issue.

What is really needed is rate relief, timely and adequate rate relief. The use of a subsidiary, project financing or related financing techniques merely postpones rate relief or obscures the need for it. During the construction phase the financial needs accumulate and massive rate increases are needed at completion to carry the investment and the cost of operation. Since there is no certainty as to the amount and timeliness of rate relief, the attendant risk of alternative financing techniques can be too large to accept.

Whether we will have sufficient coal-fired capacity or, for that matter, any form of generation lies in regulation and the willingness of regulators to allow rates to permit the attraction of capital. The rate-making process relies heavily on the historic test year and embedded capital costs. This must change. The problems lie ahead, not behind. We cannot finance incremental construction at incremental capital costs with rates reflecting yesterday's costs, especially when a final decision is reached as much as a year or two after an application is filed.

There is nothing wrong with the manner in which we financed in the past. It was and is sound and proved. Perhaps we should decrease our reliance on long-term debt to some degree. But this is only possible if our rates and earnings performance support the added common equity. Major revisions in our financing approach are not a proxy for adequate rates.

Rate making must look forward, not backward. Times have changed. The era of cheap energy has passed. The public must come to understand this if we are to continue to have a viable supply of electric power. To date, it has not.

The public is inflation-weary, and the price of electric power is set in a public forum. The public—directly and through their elected and appointed representatives—has a loud voice in rate making. That voice has said, "Keep the rates low." In part, this cry is conditioned by the sharp increase in the price of coal which has been passed on to the consumer. But that voice is sacrificing the future adequacy and reliability of electric power for short-term expediency. Is this in the public interest?

The history of electric power in this country in terms of availability and reliability of supply is unsurpassed elsewhere. The knowledge and technology to continue that record are established. Coal-fired generation is not a dream, in the laboratory, or a prototype. It is daily supplying our energy needs and can assume a growing role in the future. We also know how to finance it and all the other facilities essential to electric power supply. And the financing can be done if we all recognize that periodic rate increases must be granted on a timely basis. We must pay as we go—we cannot go now and pay later.

The rise in the cost of coal has led to heightened opposition to general rate increases needed for financing purposes. Price, as well as availability and continuity of supply, is of paramount importance to electric utilities. For coal-based utilities, coal now represents up to 70 percent of total operating expenses. The choice between nuclear and coal-fired generation rests heavily on the price of coal. Coal can be priced out—or "regulated" out—of the market.

The nation's coal producers have a major opportunity and challenge ahead. Not the least challenge will be to solve the ongoing productivity problem.

Electric power is fundamental to our life-style. Anyone who has experienced a protracted power outage knows its importance to us. Capacity reserve margins nationally are dwindling. If we allow a crisis in electric power to occur, the road back will be a long one. The social and economic consequences of such a crisis are not acceptable.

# 11 Development of Federal Coal Resources

*Frederick N. Ferguson*

Although there have been special statutes concerning the disposition of coal resources owned by the federal government since at least 1864, there really has not been much interest shown by the general public in these resources until the last few years. Now in this day when the general public has at last become conscious that we must change our ways of squandering energy, people have become aware that the United States, in its proprietary capacity, owns immense supplies of recoverable coal, mostly in the Western states. And at the same time as the general public has become aware of these great federal resources to help in the alleviation of energy shortages, it has become aware of possible damage to the environment if these resources are not developed under careful controls. Energy needs and environmental concerns, and the apparent clashes between them, have made life in the Interior Department, and particularly for lawyers, very exciting during the last six or seven years.

Much of the land in the Western states, over 400 million acres, is still public domain whose title has never passed from the United States. In other lands, some 55 million acres, the surface has been conveyed to private parties, but the coal has been reserved to the United States. Not all this land is valuable for coal, of course, but much of it is. The Bureau of Mines has estimated that the total known recoverable resources of coal in federal ownership are now over 100 billion tons, of which a little less than 50 billion is subject to surface mining and a little less than 60 billion to underground mining. Others maintain that these are very conservative estimates. Certainly, these are great resources and, if properly developed, they could contribute significantly to the fight for energy self-sufficiency. Production from federal lands has been small, but is now increasing. Even in 1973 federal production was only 12.9 million tons out of a national total of 592 million; but in calendar year 1976 production was 38.5 million tons, and for the twelve months ending on September 30, 1977, it was 50.2 million tons. The latest estimate of the recoverable federal coal resources already in the hands of federal lessees is 17.3 billion tons. Western coal is for a number of reasons increasingly attractive, not least because of its low sulfur content, and the federal resources are mostly Western. If the Department of the Interior can establish a sound coal leasing program and meet the requirements of proper environmental protection, a much greater share of the national production can, and indeed will, come from federal resources.

It is desirable to clarify the nature of the statutes governing the disposition of federal coal resources. First, the United States does not mine the federal coal itself. The method established by statute is the issuance of coal leases by the Secretary of the Interior to private parties. A coal lease grants a party the right to extract the coal upon the payment of a rental for the land and a royalty on all production. Coal deposits in public domain lands are leased under the Mineral Leasing Act of February 25, 1920;[1] coal deposits in acquired lands are subject to the Mineral Leasing Act for Acquired Lands[2] which became law in 1947. The Acquired Lands Act incorporates by reference most of the provisions of the 1920 act. Naturally, there have been amendments to the 1920 act in these 58 years since it first became law; the most important amendments were enacted on August 4, 1976, in a statute strangely named the Federal Coal Leasing Amendments Act of 1975.[3]

Under the provisions of the Mineral Leasing Act, the Secretary of the Interior is not required to issue coal leases, but he has discretion to decide whether to do so. Any coal leases which he does issue must be awarded on a basis of competitive bidding. Leases may be issued only to United States citizens, corporations, or groups of citizens or corporations. People often ask whether foreign interests may someday dominate our energy industry. By statute aliens may hold shares in American corporate lessees only if their own countries grant comparable rights to United States citizens, but there is nothing to prevent citizens of reciprocating countries from controlling American corporations which hold coal leases.

Before the Secretary offers a coal lease for bidding, he must have a comprehensive land-use plan prepared which covers the tract to be offered, and he must determine that leasing the tract is compatible with the plan. A reasonable number of tracts must be reserved and offered only to public bodies and rural electric cooperatives. Another limitation which has recently been imposed in an effort to ensure diligence is that a party who has held a federal lease without production may not obtain another lease.

There is no statutory limit on the size of a federal lease, but there is a limit on the total acreage which a person may have under lease—46,080 acres of public domain in any one state and 100,000 in the nation as a whole, and comparable amounts of acquired lands. A coal lease is for a term of twenty years and so long thereafter as coal is produced each year in commercial quantities. However, no one can merely hold a lease for twenty years; he must be producing coal within ten years. That is only one of the many spurs to activity. A lessee must submit an operation and reclamation plan within three years after the issuance of a lease, and that plan must call for the complete mining of the reserves within 40 years from the approval of the plan. Federal leases may be included in logical mining units which may include state and private land as well as federal land. Generally speaking, activities on one part of an LMU, as we call a logical mining unit, count as activities on all parts of that LMU.

An annual rental is charged on all leases. Recent leases have called for $3 an acre. The 1920 act originally set a royalty rate of not less than $.05 per ton. The Department of the Interior set higher rates, and leases issued in the last five years have had royalty set at 8 and 10 percent of the value of the coal mined, with a floor of $.40 or $.50 per ton. Now, as a result of the 1976 amendments, the royalty rate must be not less than 12.5 percent of the value of production, except that the Secretary may set a lower rate for underground coal.

Beyond these conditions which are specifically imposed on a coal lessee by statute, many of which have been added by the 1976 amendments, the Secretary of the Interior has wide discretion. Section 30 of the Mineral Leasing Act[4] has never been amended since its enactment in 1920, and it is quite my favorite section in the law. Section 30 sets out various provisions which must be included in every lease, whatever the mineral may be. These provisions include several which address the social problems of 1920, such as the 8-hour day, freedom of purchase for workers, and payment in United States money. Section 30 then requires the Secretary to include in the lease "such other provisions as he may deem necessary . . . for the protection of the interests of the United States . . . and for the safeguarding of the public welfare." It is this broad authority which has enabled the Interior Department to adjust its mineral leasing program to suit the changing conditions of the present decade. It is under this section that we have been able to include in leases provisions which give necessary protection to the environment, for example. The Department of the Interior has been able, by judicious use of this and other authority, to switch from a leasing program which in effect left to the lessee the decision of whether to produce and whether to protect the environment to a policy which has imposed on lessees both incentives and even compulsion to produce and a necessity for proper environmental protection. It was quite remarkable to see how far the Department of the Interior progressed even before the statutory amendments of 1976 in addressing the two aspects of the federal coal leasing program which have drawn the most criticism: (1) the fact that there was so little production from the 17.3 billion tons already under the lease and (2) the failure to require adequate environmental protection.

One provision in the Mineral Leasing Act which was repealed in 1976 still has a major bearing on federal coal leasing. The Mineral Leasing Act used to authorize the issuance of not only coal leases, but also coal prospecting permits. Prospecting permits could be issued where either the existence or the workability of coal resources was not known. Unlike the leases which even then were issued by competitive bidding, a prospecting permit was issued to the first qualified applicant, and, if the permittee within the prescribed term discovered coal in commercial quantities, he was entitled to a noncompetitive lease. These noncompetitive leases are commonly, but incorrectly, called *preference right leases*. Even now there are approximately 190 applications for such leases pending, which cover about 9 billion tons. These applications have become quite a matter for controversy. Although Interior believes we have no discretion to refuse to issue a noncompetitive lease to a permittee who has discovered coal in commercial quantities, environmental groups have asserted in court that the Secretary has full discretion to refuse to issue a lease. On the other hand, in 1976 the Department of the Interior issued regulations which took the position that to show coal in commercial quantities, an applicant had to meet a marketability test similar to that under the mining law and in determining whether a deposit was marketable, all the lease terms and conditions, including those for environmental protection, have to be considered. This interpretation has not pleased the industry. Finally, a recent Solicitor's Opinion M-36893 has cast doubt on the extent of resources covered by some applications. Obviously, if the resource covered by an application is drastically reduced, it may no longer be marketable. Clearly there are lots of problems involved in the disposition of these 9 billion tons of reserves.

In brief, that is how the Mineral Leasing Act provides for the disposition of federal coal resources. Unfortunately, the issuance of coal leases is not simple. The Mineral Leasing Act itself requires many time-consuming steps where informed judgment must be exercised. But we must have a land-use plan before a tract is offered. We must establish lease terms which will protect the interest of the United States and safeguard the public welfare. Interior must do many other things. Although full responsibility for the issuance of coal leases and the supervision of operations under those leases remains in Interior, the new Department of Energy has been given responsibility for issuing regulations on diligence requirements and rates of production and for fostering competition, and Interior must clear any lease terms on those subjects with the new Department of Energy.

In determining whether to offer tracts for lease and which terms to include in a lease, there are many other statutes besides the Mineral Leasing Act with which we must comply. A Department of Interior lawyer recently compiled a list of environmental statutes which apply to federal coal leasing. His list contains thirteen. Overall, there is the National Environmental Policy Act[5] which requires the preparation of an environmental impact statement for any major federal action significantly affecting the quality of the human environment. Is the issuance of a coal lease such an action? One must consider each proposed action carefully. Environmental impact statements are neither brief nor easy to write. The Department's first environmental impact statement (EIS) was written in 1970 on a proposed sale of offshore oil and gas leases. The first statement was only 35 pages long and was written between noon Friday and 4 a.m. Monday by four people, but those were innocent days and we never write statements like that today. At least two substantial volumes are usual.

The purpose of all these statutes is to prevent us from being single-minded in our approach to coal leasing. We must remember other uses of the land and the impact that coal leasing will have on the whole community and the nation. The mandate that we consider all these other factors is not inimical to active coal mining. What we are striving to achieve is balance—balance between proper development of our vast coal resources and proper protection of the environment. This is an objective with which all would agree, but different people see the balance in different ways. This country has immense resources, and they must be used and developed side by side, not by reckless exploitation of one at the expense of the others.

A natural reaction could be that all the foregoing is rather theoretical and does not reveal where we stand right now and what we are actually doing. An honest answer is that we have been working very hard and moving very slowly. In 1973 the Department stopped issuing prospecting permits which were authorized before the Mineral Leasing Act was amended in 1976 and stopped issuing leases except where the Department's short-term criteria were met. Only a few leases have been issued in the last seven years. The Department began the consideration of a new coal program. An environmental impact statement on the whole coal program was begun. This is the kind of statement which we call a "programmatic" as distinguished from a "site-specific" one, which concerns only a very specific action such as one sale of coal leases in a limited area. Litigation has resulted, and last summer in the Federal District Court in the District of Columbia, in *NRDC v. Hughes,*[6] the programmatic statement was ruled inadequate and the Department was prohibited from any coal leasing except under most restrictive conditions. Fortunately, Judge Pratt's decision does not restrict the actions which we take on existing leases, such as the approval of mining plans. However, it has so limited our authority to issue new leases that few, if any, are likely to be issued.

We have begun the formulation of a new coal leasing program and the preparation of an extensive supplement to the programmatic EIS. Before the programmatic statement can be properly written, the new program about which it will be written must be formulated. The present expectation is that a final programmatic, with all the required studies and public hearings, cannot be completed until the end of 1978. Then, and only then, would come a decision on whether to adopt the proposed coal program. If it were decided to adopt the program, the next step would be to select specific tracts for consideration for leasing. Site-specific environmental impact statements would then be required, and they would not be completed until 1980. After that, a decision could be made whether to offer any tracts for lease.

Just what factors should be considered in formulating a leasing program? An important question is whether any new leasing is needed at this time. Some people point to the reserves already leased. The figure given for the leased reserves is 17.3 billion tons. At the present rate of production, or even an increased rate of production, that could last a long time. Diligence requirements are expected to speed up production. There are about 190 preference right lease applications; these cover more than 10 billion tons. There are many questions about the preference right lease applications, but if these applications should be successful, the amount of coal under lease would be greatly increased. What about the impact of federal leasing on communities in the Western states? The Mineral Leasing Act grants revenues to states for the use of impacted areas, but these funds are granted after the impact, not in time to plan for the impacts. Interior has promulgated regulations which would allow a lessee to surrender rights in one area and to receive rights in another.[7] These regulations offer-a good opportunity to avoid the most undesirable impacts of federal coal mining. Surprisingly, both industry and environmentalists have welcomed the concept. How do we balance the pros and cons of surface mining and underground mining?

Achieving true competition for federal coal leases faces some special hurdles. Much of the federal land in the Western states is checkerboarded, because the land grants of the last century often gave railroad companies alternate sections. This pattern has often existed until the present day. Naturally, the coal resources do not conform to the ownership patterns, and the most efficient mining operation may require that private and federal resources be developed together. When a federal lease is offered for a tract adjoining one in private ownership, the owner of that private tract may be in a position superior to that of any other possible bidder. Similar problems arise where the surface belongs to a private party and the coal resources to the United States.

We have lots of questions. Perhaps, we shall be able to answer many of these questions a year from now.

## Notes

1. 30 U.S.C. §§ 187-263.
2. 30 U.S.C. §§ 351-359.
3. P.L. 94-377; 90 Stat. 1083.
4. 30 U.S.C. § 187.
5. 42 U.S.C. §§ 4321-4347.
6. Civ. Action No. 75-1749, D.D.C., Sept. 27, 1976.
7. 42 F.R. 64346 (Dec. 23, 1977).

# Part IV
# Uranium as a Source
# of Energy

# 12 Electric Power Needs of the 1980s

*Aubrey Wagner*

Three conditions would preface any discussion of electric power needs in the 1980s:

1. The need for electric power in the 1980s and beyond will increase substantially.
2. It is imperative to our national well-being—economically, socially, and from a national security standpoint—that those needs be met.
3. As things look now, there is a strong possibility that they may not be met.

Continued population growth, along with an inevitable shift from oil and gas to electricity, will ensure that the demand will increase. Because of programs to encourage energy conservation and because of the rising cost of energy, the growth rate will probably be slower than in the past. But it will still be substantial. Conservation, coal, and uranium are the keys to the energy future of this nation, at least until the end of this century.

Football coach George Allen is given to terse philosophy. In his efforts to inspire his team as they approach and enter the playoffs, he reportedly tells them, "The future is now." If we don't make it today, there will be no tomorrow.

As we consider the electric power needs of the 1980s, the only change to better state our situation is, "The future was yesterday." Energy planners must live in the future. The generating capacity which provides the electricity we are using today in this country was planned years ago by a group of people who were looking ahead to 1978 to see what the demand would be and what methods should be used to meet it. That's not an easy job. It's an art as well as a science, and an imprecise one at best.

To begin with, power planners must forecast needs ten to fifteen years in advance in view of the delays which, for one reason or another, have crept into the business of building electric generating stations today. To illustrate the difficulty of this long-range forecasting, how could one have forecast today's energy needs in the early to middle 1960s when few visualized an oil and gas shortage, when coal cost $4 a ton, or when the environmental impacts of burning fossil fuels were almost unrecognized?

In addition to normal factors involved in any long-range forecast, the electric power planner must consider such uncertainties as the licensing process,

89

site availability, various environmental factors, the cost and availability of fuel, the problems of financing such large-scale construction, the impact of increasing prices on power use, the impact of conservation programs, the effect of substituting fuels one for another, the impact of legislation which Congress may or may not pass, and a host of other factors. All this must be estimated for a period ten to fifteen years in the future.

Yet, the answers we come up with in 1978 must still be the right ones in, say, 1990, when the capacity is actually needed. Unfortunately, the forecaster's book does not have any answers in the back. Only time will tell whether we were right or wrong. And we cannot afford to be wrong by much.

When you consider the vital importance of energy, and in this case electric energy, in the daily life of every American, it seems clear that if we err, we ought to err on the side of too much capacity rather than face the destructive consequences of brownouts, blackouts, and the economic collapse that is sure to follow if we err on the other side.

A year ago the Federal Power Commission (FPC) warned that regional electricity shortages are a distinct possibility in the first five years of the 1980s unless planned power plants are allowed to begin operating and several other conditions are met. The FPC said, for example, that higher demands for electricity coupled with delays in operation of *nuclear* plants will *ensure* shortages.

The situation during early January 1978, in the face of bitter cold weather, illustrates the accuracy of this forecast. In the Tennessee Valley, for example, licensing and regulatory delays in the Sequoyah nuclear plant project have prevented its coming online even yet, although it was originally planned to begin operation in 1973. This, coupled with manufacturer's problems for the Raccoon Mountain pumped-storage project, which have delayed it for two years, left us about 4 million kilowatts short of our planned capacity for this winter. As a consequence, on Monday morning January 9, 1978 we were able to carry our load with only a razor-thin margin. An emergency outage at any one of our medium to large generating units would have required our dropping firm load to avoid losing our system.

As the months and years go by, unless schedules for planned power plants can be regained, the certainty of continued load growth will, as FPC warns, ensure shortages. In other words, while "deferral," "postponement," and "delay" sound harmless enough years before a plant is scheduled for initial operation, in reality, they plant the seeds for disaster of major proportions.

These are situations which add further to the uncertainties that face electric energy planners. For them, the problem boils down to one of doing their best to estimate the future *demand* for electricity and the *capacity* needed to supply that demand, including an adequate margin.

Let's turn first to the question of demand. There have been many forecasts made for the 1980s and the years beyond. For the nation as a whole, a half-dozen organizations close to the business have estimated that electric power loads will grow over the next ten years at annual rates varying from 5.7 to 6.4 percent. Considerable variation is estimated among different regions of the nation, ranging from 4.8 percent in the Northeast to 7.4 percent in the mid-continent area.

In the TVA region, we estimate that our loads for the next ten years will grow at an average annual rate of 5.7 percent. These figures may be compared with a national average annual growth rate of 7 percent, or perhaps a little more, in the period preceding the Arab oil embargo.

In absolute terms, rather than percentages, the nation's electric utilities are currently generating about 2 trillion kilowatthours per year. Within 10 years the annual needs of the nation are estimated to run between 3 and 3.5 trillion kilowatthours. In terms of peak loads, the annual peaks are currently running about 400 million kilowatts, and within ten years they will probably be in the neighborhood of 650 to 700 million kilowatts. On TVA's system, we supplied 118 billion kilowatthours in 1976, and we estimate this will increase to about 210 billion kilowattshours by 1986. Our peak load in 1976 was 20.6 million kilowatts and has since passed that mark. We estimate that by 1986 we will have to carry a peak of about 36 million kilowatts. All these figures amount to about a 70 or 75 percent total growth rate over a 10-year period.

It is important to recognize that these estimates of future load growth are not simply paper extrapolations of historical growth rates. They are backed up by the most precise and detailed analyses of the many segments that build up the total load comprised of residential, commercial, industrial, street lighting, and special loads.

Forecasts include both factors that will tend to increase the loads which must be carried and factors that will tend to decrease those loads. Population growth is a factor in determining loads. Most estimates project about 50 million more people in the nation by the end of this century. They will help to account for an increase of some 15 million households over the next ten years. Electric energy will be required to heat those households and to operate the water heaters, other appliances, and conveniences the American people now demand.

Over the next ten years, total employment is forecast to increase by 12 or 13 million jobs. It will take more energy to run the offices, factories, and shops where these people will work.

As gas and oil become less available, in many instances electric *energy* will substitute for them. Thus while growth in our total use of energy may slacken or even tend to level off, our use of *electricity* will continue to rise.

Increasing amounts of energy also will be required for cleaning the environment. For example, air-cleaning equipment being backfitted on existing coal-burning stations, as well as on new stations, will consume as much as 10 to 15 percent of the output of those stations. Not only does this require the addition of more capacity in the future, but it reduces the capacity already available to serve present needs.

The automobile, as we all know, is a major user of energy in this country. Granted that cars must be more efficient in the years ahead, it is certain that the American people will continue to drive automobiles. To what extent will this load be placed on the electric utility industry as our oil supplies begin to dwindle? The answer to that question can profoundly affect electric power loads. We estimate, for example, that a small, lead-acid, battery-powered electric automobile would add about 5000 kilowatthours per year to a family's annual consumption of electricity, which is about 50 percent of the present level of consumption in the average American home. More sophisticated electric cars, capable of direct use-replacement of our present gasoline-driven autos, would use considerably more electricity.

Of course, there also are factors that will tend to reduce the consumption of electricity in the years ahead. The inevitable increase in price will undoubtedly be such a factor. As prices increase, the incentive for conservation becomes ever greater. More homes are being insulated, and new homes are being built with energy conservation as a major objective. Heat pump sales are booming. The efficiency of appliances is improving, and more will be done in this area.

In addition to price as an incentive for conservation, a growing public awareness of its importance has a very substantial impact on energy use. In the TVA area, our load projections have been made—first, without taking into account either the effects of conservation or the substitution of electric energy for other energy sources. Then we have separately estimated the impact of conservation and the impact of substitution. The net effect of these two factors will be to reduce our total load ten years hence by 6.9 percent. This is the result of an estimated 10.3 percent reduction from conservation and an increase of 3.4 percent from substitution. Among residential consumers the net effect is nearly twice as great, resulting in residential loads 12.2 percent less than they would be without either conservation or substitution. We believe conservation will reduce residential loads 14.1 percent, while substitution will increase them by 1.9 percent.

Figures of this kind, of course, will vary from one section of the country to another. Conservation may have greater potential in our area where, ten years hence, about 1,605,000, or 55 percent, of our 2,918,000 homes will be electrically heated. At the same time, substitution may have a greater effect in residential areas elsewhere in the nation where electric heat will probably become more important as oil and gas become less available.

Let's turn now from demand and load forecasting to the problems of power supply. There are uncertainties and frustrations of comparable magnitude. Those responsible for providing the nation's power needs will agree unanimously that the only available technologies for new large-scale base-load generating stations are fossil-burning steam plants or light-water nuclear reactors. This will certainly be true for the 1980s and probably for the balance of this century. Since oil and gas should no longer be burned under boilers, the choice boils down to coal-fired or nuclear-powered steam plants. And this is where the fireworks start.

Those who are opposed to nuclear power say, "burn coal." Those who are concerned about air quality say, "go nuclear, or solar, or solve the problem with conservation." The fact is that we must use coal, nuclear, and conservation—all three—at the same time that we research solar energy and other forms which have promise but are not yet technologically or economically available. It should be pointed out that while solar water and space heating can save *energy* in the years ahead, it will *not*, unless arrangements (not yet developed) can be made, significantly reduce the *capacity* needs of most electric power systems. For example, the bitter cold in January 1978 followed a long period of cloudy, rainy weather. Any reasonable amount of stored heat would have been used up. The backup system—in most cases, electric—would take over, adding to the system load right on peak.

In addition to having enough capacity to carry the actual peak loads as they arise, utilities must have reserve capacity to allow for a wide variety of contingencies such as routine maintenance, emergency outages resulting from mechanical failures of one kind or another, drought or flood conditions where hydro energy is involved, problems associated with extremely cold or extremely hot weather, and others.

It is currently popular among critics of the utility industry to say that we have built too much capacity, that our current margin nationwide of about 30 percent is too much. Yet, this very week in many parts of the nation virtually every usable generator is running. Voltage reductions are common, interruptible loads have been cut off, and power is moving in interconnections to meet emergencies. Everyone is scraping the bottom of the barrel. Even so, some firm load curtailments have been necessary in some areas. The kibitzers may say that we are overinstalled. But those who are sitting at the table vigorously disagree. And the stakes are very high!

Nationwide, during the past five years, our annual peak load has grown by some 70 million kilowatts. In the next five years, it is estimated to increase by about 130 million additional kilowatts—an *increase* of more than 85 percent over the past five years' growth. To meet this load growth, the nation's capability at peak increased about 145 million kilowatts in the five years past, but will increase only about 155 million kilowatts in the *next* five years. This is a *decrease* of about 20 percent. In other words, in the next five years we will have to meet 85 percent more load increase with 20 percent less capacity increase.

It is too late already to do anything about capacity for the next five years, and at that our current margin of somewhat more than 30 percent is apparently being sorely taxed. Five years hence that margin will drop to a little over 20 percent and even lower in the years ahead.

These are some of the reasons why there is considerable doubt as to whether the nation's vital electric loads can be successfully met in the 1980s. These are some of the reasons why the National Electric Reliability Council, as well as the Federal Power Commission, has forecast trouble ahead.

When we consider the vital role that energy—and specifically, electricity—

plays in the life of this nation, we must recognize the time has come to put a stop to delaying tactics, deferrals, postponements, and "further studies" of energy situations. We must, of course, proceed with care and caution. But we also must get the power train back on schedule so that the capacity will be there to meet the loads when they arise.

Energy is far too precious a commodity to waste. And conservation is important, almost beyond any ability to state it. Those of us who are electric power suppliers have an overriding responsibility to provide all the electric energy the people of this country really *need,* at prices they can afford—consistent with maintaining a clean environment. This we must do. It can be done if we will but bring some balance into our judgments and decision making and then get on with the job—together.

# 13 Alternative Nuclear Fuel Cycles

*Charles Till*

## Introduction

In this chapter we attempt to summarize the implications of fuel-cycle selection by criteria based on proliferation grounds on achievable nuclear electric capacities within a limited natural resource base. The question is, How far can uranium supplies be stretched if criteria of this kind are used to define admissible reactor cycles? The question is a simple one. The problem in answering it lies in the very number of reactor types, fuel cycles, and deployment permutations, as well as the widely varying opinion currently held on both the magnitude of the uranium resource base and the amount and time span of nuclear power that may represent an adequate objective. For timely decisions in the nuclear power development program, answers are needed in the form of dates:

The dates when resource limitations may become binding

The corresponding dates by which decisions to implement new reactor types must be made

The dates by which developmental decisions must be made

Many of the calculations required to estimate these dates are not difficult, particularly once the fuel flows for the various reactor types have been computed. The problem is to assemble the information in a way that it can be digested. Further, while recognizing that answers to many of the questions cannot be given with certainty, our aim is to present the data in a way that at least shows the possible ranges.

Four possible policy constraints define different reactor deployment strategies:

1. No fuel reprocessing, converter reactors only, once-through cycle
2. Fuel reprocessing, converter reactors only
3. Converter reactor/fast-breeder reactor limited deployment
4. Converter reactor/fast-breeder reactor with limitations on fuel characteristics but not on deployment

Fuel utilization characteristics for the light-water reactor (LWR) and for three advanced converter reactors for a variety of fueling options are shown in figure 13-1. The breeding performance of liquid-metal fast-breeder reactors (LMFBRs) also under a variety of fueling options is shown in figure 13-2.

Future nuclear electric demand and the uranium resource base can be specified only in terms of ranges. For demand, we take 380 $Gw_e$ in the year 2000, the recent estimate given in congressional testimony by Department of Energy (DoE) Secretary Schlesinger, with further growth at rates of 10, 20, and 30 gigawatts (electric) per year thereafter. For a range of uranium resource base, we selected 2.4 million standard tons (ST) $U_3O_8$, the ERDA "prudent planning estimate," and 4.3 million ST $U_3O_8$, the most recent ERDA estimate of possible high-grade ores, as reasonable bounding estimates.

**Converter-Only Strategies**

Figures 13-3 and 13-4 present the full range of possibilities for converter reactors that are well enough developed that reasonably accurate estimates can be made

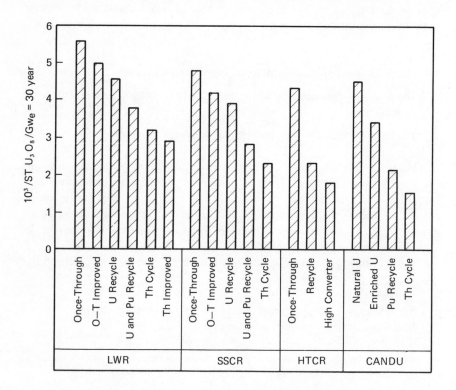

**Figure 13-1.** Summary of Converter Reactors' Fuel Utilization Characteristics (70 Percent Capacity Factor, 0.2 Percent Tails Assay).

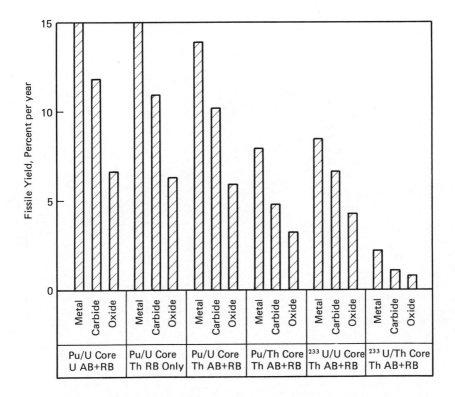

**Figure 13-2.** Summary of LMFBR Breeding Performance (70 Percent Capacity Factor).

of their fuel utilizations. These figures require some explanation. First, they show the final year that reactors can be brought on-line with ensured 30-year forward commitments, for each of the two assumed resource bases. That year is labeled the converter phase-out year and is best understood by reference to figure 13-5. Figure 13-5 displays a calculation for one combination of demand and resource base. The circled points show the phase-out year for two candidate reactors. In the calculation, reactors are brought on-line at a rate just equal to the demand plus the retirement rate up to the year that 30-year forward requirements can no longer be met. The capacity on-line then drops according to the retirement rate until in the final year the assumed resource base is exactly used up.

Figures 13-3 and 13-4 can be viewed as plots of many such calculations. The vertical scale quantifies the system average lifetime $U_3O_8$ requirement. This needs a word of explanation as well. This quantity is the average $U_3O_8$ consumption per gigawatt (electric) over the duration of the nuclear enterprise until all the resource base is consumed. The need to define such a quantity arises because of our wish to make figures 13-3 and 13-4 applicable to a range of

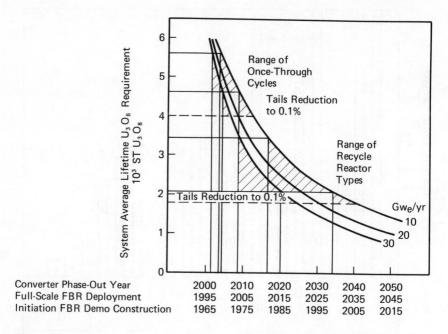

**Figure 13-3.** Effects of Converter Reactor Improvements on Reactor Program Timing (Assumed Resource Base 2.4 Million ST $U_3O_8$).

different reactor types. The $U_3O_8$ requirements to the year 2000 are largely dominated by the LWR. Thus, for example, when a new reactor type with improved fuel utilization is brought on-line, the system average lifetime requirement will be a weighted average of the characteristics of the LWR and the new type. To characterize the ranges, figures 13-3 and 13-4 assume the introduction of advanced converter reactor in 1995, and the lower bound for each class of fuel cycle is defined by full-scale introduction in that year of the most fuel-efficient advanced converter reactor on its most fuel-efficient cycle. The shaded triangles give the maximum incentive to develop improvements of the type shown in terms of the extension they could provide in the duration of the nuclear enterprise. For example, the upper two triangles characterize the incentive for introduction of advanced converter-reactor types, if once-through cycles only are allowed. Reduction in the tails assay to 0.1 percent has also been suggested, and the incentive for that is also shown. The range of recycle effects is defined on the high side by the LWR on $^{233}U$/Th recycle and on the low side by the advanced converter reactor, in this case, the CANDU reactor on self-generated $^{233}U$/Th recycle. The effects of tails reduction again are shown.

If the fast-breeder reactor (FBR) is brought in to forestall nuclear electric

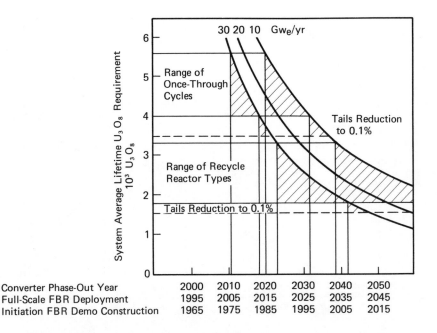

**Figure 13-4.** Effects of Converter Reactor Improvements on Reactor Program Timing (Assumed Resource Base 4.3 Million ST $U_3O_8$).

phase-out, the second scale on the ordinate shows the year full-scale fast-breeder reactor deployment is required. The upper scale assumed plutonium recycle in the recycle converter case, so the second scale is offset about five years to provide the plutonium requirements for the FBR initial inventories. This five-year offset is not needed for the once-through cycles, and with fast-breeder reactor follow-on from once-through cycles this adjustment should be made. Also shown is the estimated initiation date for FBR demonstration plant construction, in order to provide the lead time required for introduction of breeder reactors on the date shown at a rate of 20 to 60 gigawatts per year as appropriate to the assumed demand and retirement rates.

### Converter/Fast-Breeder Strategies

Turning now to deployment scenarios that include the fast-breeder reactor, we have calculated the maximum capacity possible for both the 2.4 and 4.3 million ST resource bases. Figures 13-6 to 13-10 summarize information on deployment scenarios based on LWRs and oxide-fueled LMFBRs. To 1995, LWRs on

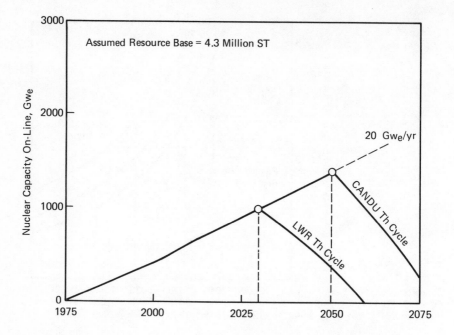

**Figure 13-5.** Illustration of Converter Reactor Phase-Out Year.

uranium recycle only are assumed, and a total of 20 gigawatts (electric) for the five-year period 1995 to 2000 is taken as an introduction constraint on the LMFBR. The LMFBR may utilize thorium, partially or fully, as its fertile material. The bred $^{233}U$ is used to fuel denatured LWRs. The capacity after the year 2000 is allowed to vary to maximize the total energy produced. (Reference 1 contains a more complete exposition of the methodology and a more detailed presentation of these symbiotic deployment strategies.)

Denatured reactor cycles can be thought of as two types: First, those based on converter reactors operating with thorium as the primary fertile material; and second, those using isotopic mixtures of uranium only. In figures 13-6 and 13-7 the converter is of the first kind, and the $^{233}U$ needs some isotopic dilution with $^{238}U$. This calculation assumed 12 percent $^{233}U$; the remainder of the fuel is thorium. The figure shows the effect on nuclear capacity of increasing amounts of thorium in the LMFBR. The top curve shows the case with no thorium at all and thus gives the growth potential for the mixed oxide LMFBR alone. Below that in order of decreasing capacity potential are the cases with thorium substituted in the radial blanket only, the case with both axial and radial thorium blankets, the case with a heterogeneous core design that includes internal thorium blankets; finally, the lowest curve gives the effect of complete substitution of thorium for $^{238}U$ in both core and blanket.

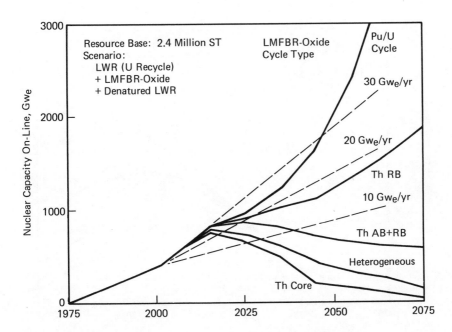

**Figure 13-6.** Comparison of Nuclear Power Growth Potential for Various LMFBR Cycle Types (Resource Base 2.4 Million ST).

The figure shows the deleterious effect on growth potential of the thorium addition. Of the thorium cases only the thorium radial blanketed LMFBR case shows long-term growth potential. It has an important disadvantage, however. The other important factor in assessing denatured converter/LMFBR-$^{233}$U producer cycles is the magnitude of the ratio of the two reactor types. The higher the ratio, the greater the freedom to install capacity outside protected areas. Figure 13-8 gives this ratio as a function of time for the four thorium-based cases. It shows the "thorium radial blanket only" case to be poorest from this point of view, and, as might be expected, the ratio is inversely proportional to the power growth potential. The higher the growth sought, the lower the fraction of converters supportable.

Figures 13-6 and 13-7 dealt with the symbiotic cycle in which thorium is the basic fertile material for reactors outside the fuel center. The other alternative makes $^{238}$U the fertile material of preference for reactors outside the fuel-cycle center. This allows the fresh fuel to be isotopic mixtures purely of $^{233}$U and $^{238}$U, and the plutonium content in the spent fuel is then recycled back into the fast-breeder reactors in protected areas. In this cycle, therefore, isotopic denaturing of the fresh fuel outside the fuel-cycle center is maintained, while advantage is taken of the superior breeding properties of plutonium for the

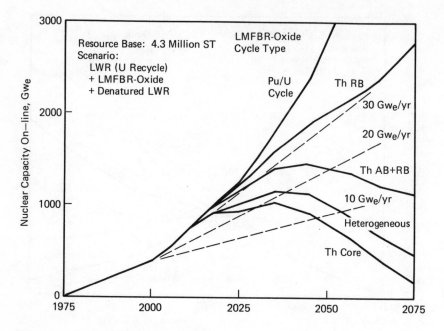

**Figure 13-7.** Comparison of Nuclear Power Growth Potential for Various LMFBR Cycle Types (Resource Base 4.3 Million ST).

breeders in a protected area. Figures 13-9 and 13-10 show typical results. The upper curve shows the growth potential for an all-LMFBR scenario, with plutonium-fueled LMFBRs inside the center and $^{233}U/^{238}U$-fueled LMFBRs outside. The lower curve shows the situation for the LWR, and the CANDU case lies in between.

## Implications of Isotopic Nonproliferation Criteria

Now let us examine the implications of isotopic nonproliferation criteria on acceptable fuel cycles and resource utilization.

(1) The case of no fuel reprocessing, converter reactors only, once-through cycle implies that there is no acceptable criterion other than adherence to the present situation. Figures 13-3 and 13-4 show that for the low resource base, the nuclear capacity will start to phase out in the decade after the year 2000, whether demand is low or high. For the 4.3 million ST resource base, the phase-out will start in the next decade for the high demand case, and is postponed another decade for low demand.

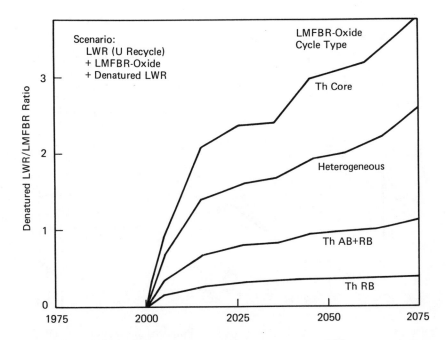

**Figure 13-8.** Comparison of Denatured LWR to LMFBR–$^{233}$U Producer Ratio for Various LMFBR Thorium Cycle Types.

(2) The case of fuel processing allowed, converter reactor only, depends on the proliferation protection criteria used. Isotopic separation of $^{233}$U from $^{238}$U is easier than $^{235}$U from $^{238}$U because of the greater isotopic mass differential, and the fast spectrum critical mass is also considerably smaller for $^{233}$U than for $^{235}$U. These two factors combine to suggest that separation of a weapons mass is perhaps a factor of 10 easier for $^{233}$U than for $^{235}$U. Analogy to an allowable 20 percent $^{235}$U in $^{238}$U would put the allowable $^{233}$U content in the 2 percent range. For simplicity we take 4 percent $^{233}$U in $^{238}$U as allowable, numerically the same as the present LWR $^{235}$U fissile content, and examine the consequences. Limiting fresh subassemblies to less than 4 weight percent $^{233}$U in $^{238}$U and allowing no plutonium in fresh subassemblies eliminates this option, unless radiation or other means for protection of fresh fuel are acceptable and can be met. If both depend on radioactivity for their nonproliferation acceptability, $^{233}$U and plutonium are essentially equivalent. $^{233}$U contents well above 4 percent in $^{238}$U must be acceptable before a denatured thorium cycle becomes feasible. If 12 percent $^{233}$U in $^{238}$U is acceptable, as has been suggested, the range of the best fuel utilizations for converter recycle is defined by the LWR on thorium recycle at the high end and

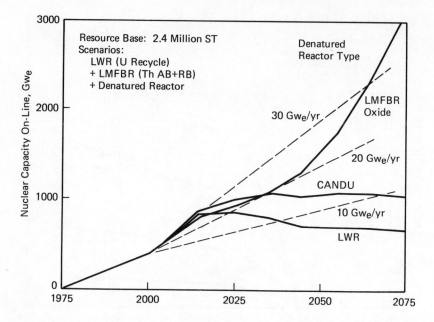

**Figure 13-9.** Comparison of Nuclear Growth Potential for Various $^{233}U/^{238}U$ Fueled Denatured Reactor Types (Resource Base 2.4 Million ST).

CANDU recycle on the low end. For the case of a low resource base, phase-out comes sometime during the second decade after the year 2000, and for the high-resource-base case, phase-out is postponed a further two decades.

(3) The case of converter reactor/fast-breeder reactor limited deployment also depends on the criteria used. If $^{233}U$ contents are limited to less than 4 percent in $^{238}U$ in fresh fuel, the only allowable cycle has plutonium fissile/part-thorium fertile breeders inside protected areas, supplying converter reactors outside that require enrichments of 4 percent or less $^{233}U$ in $^{238}U$. These could well be LWRs on a cycle very similar to the current cycle except that the spent fuel is brought back to the protected area for plutonium reprocessing for use in the breeder. CANDU reactors would be equally acceptable under this criterion and would give better overall growth characteristics. The overall system growth potential is about the same in this case as it is for the thorium-based converter reactor denatured-fuel cycles. Essentially, the loss in conversion efficiency in the converter reactor through the use of $^{238}U$ as the fertile material instead of thorium is offset by the increased plutonium availability for the breeder. This system can be characterized generally as self-sustaining over a long period.

If 12 percent $^{233}U$ in $^{238}U$ is acceptable as denatured fuel, fast-breeder

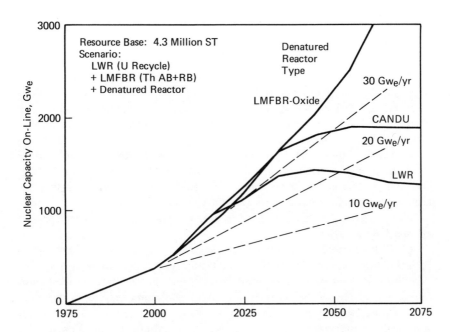

**Figure 13-10.** Comparison of Nuclear Power Growth Potential for Various $^{233}U/^{238}U$ Fueled Denatured Reactor Types (Resource Base 4.3 Million ST).

reactors fueled with $^{233}U/^{238}U$ can, if desired, also be used outside the fuel center instead of thermal converters. Figure 13-9 shows the good growth potential for this cycle.

On the other hand, if $^{233}U$ contents in excess of 4 weight percent in $^{238}U$ are unacceptable as fresh fuel, except in the presence of radioactivity, the distinction between $^{233}U$ above 4 percent enrichment and mixed plutonium/ uranium becomes less apparent, and the advantage of fueling LMFBRs outside protected areas with $^{233}U/^{238}U$ is less obvious, since either $^{233}U$- or plutonium-bearing fuels for the fast-breeder reactors could be given equal radiation protection.

(4) Now consider the case of converter reactor/fast-breeder reactor with limitations on fuel characteristics but not on deployment. Limitations on fuel characteristics would be in terms of maximum fissile fractions and minimum radiation levels when outside protected areas. Both fissile uranium and plutonium would require radiation protection. Plutonium is preferred in this case because of the superior growth potential for a pure plutonium fuel fast-breeder reactor economy.

**Conclusions**

In summary, we make the following observations.

(1) Provisional criteria for assessment of reactors and fuel cycles for proliferation risk are required. These criteria should embody present-day perceptions of acceptable risk and should not preclude future developments. If the criteria are defined in terms of levels of tangible physical phenomena, such as radiation level, minimum isotopic dilutions, and chemical form, a wide range of nuclear options is possible.

(2) On the other hand, if the current de facto criteria are rigidly adhered to, i.e., once-through cycles only are allowed, then nuclear electric probably faces phase-out starting in the first or second decade after the year 2000. Further, the incentive for development of improved once-through reactor types is not great, providing an extension of only a few years, and the same is true of developments to lower the tails assay.

(3) Criteria based on ease of isotopic separation that limit $^{233}U$ contents in $^{238}U$ to 4 percent or less eliminate converter-reactor recycle with nonradioactive fresh fuel subassemblies, unless other protective measures are acceptable. Thorium-based recycle becomes equivalent to plutonium recycle, in that both could require radioactive subassemblies or some other form of protection for fresh feed. In the thorium case, however, some additional protection may be assigned to the isotopic barrier even if $^{233}U$ contents are allowed to increase above 4 percent, say into the 12 percent range. In this range, converter-reactor thorium/$^{233}U$ recycle becomes feasible. The extension of resource usage is of the order of two decades.

(4) If plutonium breeders are allowed in protected areas, a range of possibilities opens up. With thorium blankets, they could feed LWRs with 4 percent $^{233}U$ in $^{238}U$, on a cycle very similar to the present LWR cycle. The LWR fresh fuel need not be made radioactive. Also, however, if radiation denaturing is acceptable for feed subassemblies and higher $^{233}U$ contents are allowed in association with radiation, the distinction between $^{233}U$ and plutonium fueling tends to diminish. If 12 percent $^{233}U$ in $^{238}$ is considered superior to plutonium outside fuel centers, good growth potential can be maintained with this feed for fast-breeder reactors outside the center. On the other hand, if the distinction between $^{233}U$ and plutonium is removed entirely, through the use of radiation criteria on fuel feed, bringing thorium-based cycles into the picture increases the practical difficulties and reduces the overall nuclear growth.

Finally, we observe that the need for an FBR component in any reactor deployment scenario comes back to its capacity for growth and the flexibility it gives to join with other reactor types in symbiotic deployment strategies that may satisfy external criteria, such as criteria for nonproliferation goals, and still maintain some growth potential. The highest growth potential is exhibited by an all-fast breeder economy, but the breeding potential of the fast breeder allows compromises from that if it is desirable to meet other goals. The need for the breeder as a component of the reactor economy is likely to remain, and the perception of the urgency for it seems likely to sharpen as resource usage continues.

### Reference

1. Y.I. Chang, C.E. Till, R.R. Rudolph, J.R. Deen, and M.J. King, "Alternative Fuel Cycle Options: Performance Characteristics and Impact on Nuclear Power Growth Potential," ANL-77-70 and RSS-TM-4 (1977).

# 14

# The Case for the Plutonium Breeder

## John Simpson

Participants in the development of nuclear energy over the past three decades have a feeling of nostalgia for the "good old days." They were good old days because we were fighting then real battles—the scientific, the engineering, the production, and the economic battles. We saw the task that needed doing and moved ahead with the full cooperation of the administration, the Congress, the industry, and the people. We were perceived as doing something constructive, not as being in league with the devil for the destruction of mankind. And we won.

Today there is a viable nuclear option because of what was done in those first decades. If today's climate had prevailed then, we could not have even started, much less succeeded with, the nuclear submarine. The Polaris submarine fleet is today seen as one of our most effective defenses, yet the *Nautilus* was built without knowing all the possible environmental effects, without proving it would be cost-effective. In fact, missiles such as Polaris had not then even been conceived. Central station nuclear power is today an option because our government and industry took the lead and moved ahead based on judgment that nuclear power would be needed, that it would be economic, that it would be safe.

Today that judgment has been confirmed. Clearly there is an energy supply problem. The electric utility industry sees nuclear as the economic choice in almost all cases, despite the unreasonable and costly delays and other harrassment it encounters. An overwhelming majority of engineering and scientific opinion holds that nuclear energy is adequately safe and very probably will cause the least damage to both people and the environment of any option, including doing without energy. A statement of the NAACP in a recent issue of the *Wall Street Journal* addresses this point: "We cannot accept the notion that our people are best served by a policy based on the inevitability of an energy shortage and the need to allocate an ever-diminishing supply among competing interests."

Against this background, let us assess where we stand today. There are over 200 million kilowatts of nuclear power in operation or on order or under construction. Over 10 percent of the nation's electricity comes from the atom today, and, in some of our major population centers, more than half. This 200 million kilowatts is more than the total generating capacity of the United States about twenty years ago. In the more than three decades of nuclear energy, there has not been a single death that can be attributed to radiation from a central station nuclear plant or to the research and engineering laboratories supporting the industry—truly a remarkable record.

Yet, despite this and the tens of thousands of our most knowledgeable engineers and scientists and countless reports and studies showing that nuclear energy is adequately safe, with not a single technical person who has ever been responsible for supplying energy feeling it is unsafe, despite all this, nuclear energy is beleaguered as never before. What is the message that must be gotten across to the President, to Congress, and to the people? We have reached a plateau, and there is rather general agreement, despite the few who don't go along, that the light-water reactors are a real factor in the energy supply and will remain so, but by themselves they can provide energy for only a few decades. What, then, is the answer?

1. The breeder reactor, together with the light-water reactors, can provide an almost limitless energy supply.
2. No other fuel cycle can give either the maximum capacity or the total energy needed.
3. The demand—the need—will be great.
4. The energy required for the United States and the rest of the world cannot be provided on any rational basis without the breeder reactor. The LMFBR will be safe and will cause less damage to the environment than any other available option.

All breeder reactors create fissionable material, and there is no isotope which can be transformed from a fertile material to a fissionable material and yet remain the same element. Therefore, all breeder reactors as well as all light-water reactors (LWRs) contain material that can be used for weapons. The denial of the use of the LMFBR will probably be counterproductive with respect to the containment of proliferation.

With the LWR on a once-through cycle, the peak capacity in the United States would be 600 gigawatts, and by 2025 it would be reduced to 400 gigawatts. All LWRs would be phased out by 2040. In this case, 3.7 million tons is used as the resource base. With the addition of the uranium and plutonium recycle, we can reach a peak of 850 gigawatts, which would go down to 600 by 2025 and would phase out in 2050. If we use LWRs to 2000 and then CANDUs, the situation is somewhat better, but still far from satisfactory.

The peak would be about 1000 gigawatts, and this could be maintained until 2025; phase-out would occur about 2070. With all these cycles, the total energy available from the uranium would be only a small fraction of that potentially available. With the current liquid-metal fast-breeder reactor (LMFBR) cycle, the peak capacity and the total energy would be unlimited for all practical purposes.

The consumption of energy in the United States will double to about 150 quads by the year 2000, according to opinions expressed by Westinghouse and taking into consideration the estimates of ERDA, the Edison Electric Institute, and other sources. With even as low a growth rate as about 2 percent, consumption will reach almost 200 quads by 2025. This is the equivalent of about 34 billion barrels of oil per year, but in 2025 the maximum United States oil production per year is likely to be only about 4 billion barrels a year, and the total world production only about 25 or 30.

If all the difference is made up by coal, we would need almost 10 billion tons a year. Some energy will obviously come from solar, hydro, geothermal, and others; but all these sources together will be small compared to 10 billion tons of coal. It would be unlikely that we could within economic reason mine that much coal, and if we could, it would be environmentally unacceptable.

Clearly, then, another source is required to meet the need—and nuclear energy, including the breeder, is the only such source available today or likely to be available in time. Conservation will be required, and a significant amount of conservation is assumed in all these projections.

Even with the breeder we have no way to get the quantity of energy required to maintain anything like the style of living to which we have become accustomed. With energy there is no such thing as overdrawing a bank account. The economy will adjust to the price and the quantity of energy available, and we will not like the result. The picture for the rest of the world is far worse. Of all the world's coal 90 percent is in the United States, Western Europe, and the Sino-Soviet bloc. So coal is no answer for the Third World. None of those areas that have the coal would be willing or able to export it. The oil supply of the world will meet only a small fraction of the world's needs, and thus—even more than in the United States—the breeder is needed by the rest of the world.

We cannot meet our energy needs with domestic sources, so in the years to come we will need to import all the oil that we can. This, of course, will make oil less available and more costly for the rest of the world. This brings us to possibly the gravest danger facing the United States today. The oil fields of the Middle East and the delivery chain can be disrupted almost indefinitely by about 1000 men from bordering countries such as the USSR, and there is no way the United States can prevent or stop such action. Clearly the dependence on Middle East oil is a threat to the economy and perhaps even to the safety of the United States.

With reference to the safety question, there is ample and credible evidence that the liquid-metal fast-breeder reactors can be made adequately safe, even safer than light-water reactors. Two particular characteristics make this possible: the large negative Doppler coefficient and the low operating pressure.

There can be no technical fix to the proliferation problem. Any even moderately industrialized nation can make a nuclear weapon in any number of ways that would be easier and cheaper than diverting the plutonium from an

LMFBR or a reprocessing plant. The difference in time to make a weapon from an LMFBR fuel element is but a few days to a few weeks longer than from the output from a reprocessing plant. Even more important than time is the difficulty of protecting the large number of spent fuel elements, particularly if there is to be no reprocessing and they then have no economic value. One way of reducing proliferation potential is to have secure centers in which all parts of the fuel cycle using bomb-level enriched material are located.

An energy-starved Third World, denied the benefits of nuclear energy, is no way to reduce the proliferation potential. If the United States is to have a major influence on the development of nuclear energy with respect to proliferation, then the United States must take part in all elements of the development such as the LMFBR program. The current United States posture is counterproductive with respect to containing proliferation.

Many people are willing to agree that the LMFBR may well be needed—later—so let's wait a while before constructing a demonstration plant. Unfortunately, we cannot afford to wait. If the demonstration plant is canceled now, we probably cannot have another demonstration plant in operation before the early 1990s. Remember that the Clinch River reactor program began in about 1968. The contractors were selected in 1972. The contract for those contractors was not signed until a year later. And now, four more years have passed, and construction has not yet begun.

Even a duplicate of a light-water reactor already licensed requires as much as twelve years from decision to build until commercial operation. The second generation would come about seven to ten years later, and there would only be about thirty reactors in operation in 2010. If we have no more uranium than the current best estimate of the most probable amount, or 1.8 million tons (and this is slightly earlier than the 2.3 or 2.4) and we have about 400 gigawatts of LWRs in operation in the year 2000, we are already too late by about five years. If we have 2.4 million tons, we are about on schedule—and only if we have far more than 4 million tons of uranium could we afford to stop the CRBR (the Clinch River program)—start another demonstration program later.

Almost all such major development programs such as the breeder program have been several years late in their schedules, and few programs have ever faced the institutional problems we face here. But do we need a demonstration plant? Why not go right to building a much larger plant and only carry out a research program in the interim? First, the size of components involved in the Clinch River plant is about as large as it would be wise to tackle in a first plant. A larger plant, without prior experience, would take much longer to construct—would be more costly to modify, as will undoubtedly be required—and the failure of even a single component could result in a serious delay in getting overall system operating data that would be vital before going on to additional plants.

No demonstration reactor has ever had a serious reactor physics problem, and even few nuclear core problems of any kind. But most have had numerous component and systems problems. These problems usually do not come to light in a research program. The problems arise because of the interaction of all the parts of the system and the control under the ambient conditions of both normal and transient operations.

Is the Clinch River plant obsolete, as some suggest? Only insofar as it can be said that every reactor that was ever built was obsolete before it was completed. Further, the newer the technology—and therefore the more rapid the technological advance—the more obsolete a reactor will be when completed. If a new LMFBR is started now, it will also be obsolete before it is completed. And when the Clinch River breeder operates, the lessons learned will make the new reactor even more obsolete. The Clinch River reactor can be fueled with $^{235}U$ and will contribute no more to proliferation than would a light-water reactor.

We cannot afford to depend, either, on foreign breeder technology, although it would be wise to cooperate with other countries and share the technology. We cannot depend on the timing or the objectives of other nations' programs. The other nations do not now and are not likely to meet the United States licensing requirements. There would be an adverse impact on both the balance of payments and our energy self-sufficiency goals. And, most importantly, our ability to contain nuclear weapons proliferation would be weakened or eliminated. The cost of the LMFBR program is small compared with the benefits, and almost vanishingly small compared to the gross national product over the same time period—being less than 0.1 percent.

Clearly it makes no sense to stop the LMFBR program in the guise of reducing proliferation, only to use the same technology with the same inherent proliferation potential from other countries, while giving up United States control or even United States influence.

The final summary is very simple. The demand will be there. Our way of life will be changed if the demand is not met—and there is no conceivable way in the long term that it can be met without nuclear energy, including the liquid-metal breeder reactor. And there may well be no way in the intermediate term, in the mid-1980s, that our energy requirement can be met.

# 15 Disposal of High-Level Nuclear Wastes

*Fred Donath*

Others have addressed the problems that lead to the inevitable conclusion that nuclear energy must be a major contributor to our energy needs in the immediate future if we're to maintain an energy economy likely to be acceptable to the general populace. Associated with the benefits of nuclear energy, however, comes a penalty. We must dispose of the waste generated by nuclear reactors. This problem is a burning issue of the day. No reasonable person would advocate the expansion and development of nuclear energy if there were no safe means of disposal of the wastes thus produced.

This presentation will address two basic questions that pertain to the problem. First, what is the nature of the problem? Specifically, what characteristics of the waste are of most concern to us and what volumes are involved? Second, what alternatives are there for the safe disposal of this waste?

The first question concerns the types of waste with which we are dealing and where they come from. In point of fact, the type of waste depends in part on whether we are permitted reprocessing or whether we'll be dealing with the storage of spent-fuel rods taken directly from the reactor.

In Figure 15-1 we see that from the reactor there are two possible mainstreams at the "back end" of the fuel cycle. One is to reprocess, in which case we have either the plutonium recycled or uranium recycled, or a combination, which gives rise to high-level waste that must then be disposed of in some way. It would go into a federal depository. The second alternative is to store the fuel rods themselves for an indefinite period or throw these away. In the latter instance, they would also have to go into a federal depository.

Eventually we will have to reprocess the spent fuel, and the comments in this presentation will be directed at the disposal of high-level radioactive wastes.

What volumes of radioactive waste are generated by commercial nuclear reactors? That depends very much upon the assumption one makes about energy growth in the future. Table 15-1 gives some figures that provide a perspective on the problem. Although other types of wastes are produced, we should focus on the volume and activity of the high-level wastes. The high-level wastes are of greatest concern to most individuals. Note that the volume of this waste is relatively insignificant compared with that of other wastes generated, but that virtually all the activity (i.e., radioactivity) is in the high-level waste.

There are two basic characteristics of this high-level waste that are of concern. The first, which is the radioactivity, already has been identified. In

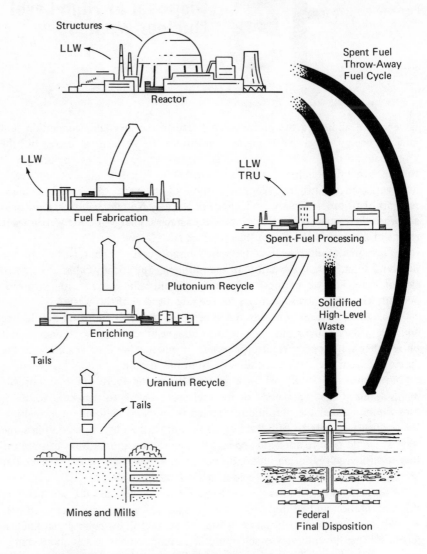

Source: NUREG-0354, U.S. Nuclear Regulatory Commission.
**Figure 15-1**. Alternatives for Back End of Nuclear Fuel Cycle.

figure 15-2 we see represented, in terms of an ingestion hazard, the relative hazard of this radioactivity over time. The ingestion hazard is the amount of water it would take to dilute this waste to the maximum permissible concentration for current drinking-water standards. It is one means of representing the health hazard of this waste.

**Table 15-1**
**Fuel-Cycle Wastes Projected for the Year 2000[a]**

|  | Annual Generation | | Total Accumulated Inventory | |
|---|---|---|---|---|
|  | Volume $(10^3 ft^3)$ | Activity $(M.in.^3)$ | Volume $(10^3 ft^3)$ | Activity $(M.in.^3)$ |
| HLW (solid) | 50 | 60,000 | 471 | 75,000 |
| CLAD | 54.0 | 250 | 450 | 900 |
| TRU | 700 | 5.5 | 5,400 | 24.2 |
| LLW | 54,000 | 1.0 | 330,000 | 4.0 |
| Gases | 2,200 | 23.0 | 21,000 | 1,500 |
| Tails | 2,400,000 | 0.8 | 31,000,000 | 9.9 |

[a] Assumes Pu recycle.

The activity results mostly from the fission products during the first 1000 years. These have largely decayed away by the end of that time. After 1000 years, ingestion hazard drops off significantly and reflects plutonium in the waste and, subsequently, various daughter products. The ingestion hazard remains at about the same level for over 1 million years.

It could be pointed out that the ingestion hazard falls below that of typical natural uranium ore after about 1000 years—i.e., the hazard of these radioactive wastes would be less, by this measure, than that of a natural uranium ore, just to give some perspective on this matter.

The second characteristic of high-level radioactive waste that is of great concern is the heat produced by it. This concern relates to the security of the isolation, as our concept of isolation today consists of emplacing the waste in a conventional mined cavity. Figure 15-3 shows a generalized plot of the radioactive heat decay from the time at which fuel rods were removed from the reactor. It is clear that the radioactive heat dies off very rapidly, such that if one were to store this material for a few years, there would be much less of a problem either in reprocessing or in emplacing it in a repository. The problems associated with heat generation in a repository will be discussed later.

So far, we have noted briefly the general nature of the radioactive waste problem. Now, what are the possible options for isolating this high-level waste? As seen in table 15-2, these can be separated into two basic categories. First, there is the actual elimination of this waste from earth, either ejection of the waste by rockets or transmuting the isotopes of concern into nonhazardous isotopes. The second major category covers various means of geologic isolation, disposal in the seabed (i.e., some place in the ocean floor), ice sheet isolation (e.g., in a continental ice sheet as in Antarctica), or disposal in deep continental geologic formations.

Both ejection by rocket and transmutation—i.e., actual elimination from earth—would require extremely efficient partitioning or separation of the isotopes of concern, primarily the actinides, from the bulk of the waste. As

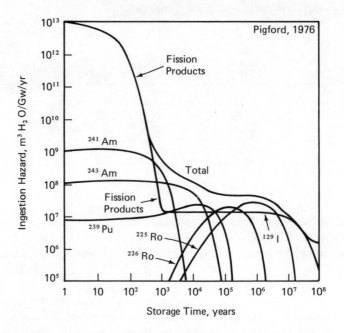

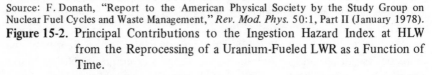

Storage Time, years

Source: F. Donath, "Report to the American Physical Society by the Study Group on Nuclear Fuel Cycles and Waste Management," *Rev. Mod. Phys.* 50:1, Part II (January 1978).
**Figure 15-2.** Principal Contributions to the Ingestion Hazard Index at HLW from the Reprocessing of a Uranium-Fueled LWR as a Function of Time.

pointed out earlier, the fission products will decay away in 1000 years; thus concern is primarily with the very long-lived radioactive isotopes. Most people would agree that we can isolate the waste without difficulty for periods up to 1000 years.

A more serious problem is to isolate wastes for much longer periods, especially if one uses the rule of thumb commonly applied, namely, that we isolate for ten or twenty half-lives. Over a period of ten half-lives the radioactivity would decay to 0.001 of the initial activity. Twenty half-lives would reduce the activity to 0.000001 of the initial.

For fission products, which have half-lives of about thirty years, we'd have to isolate these for a period of about 600 years by the above criterion. For isotopes like $^{239}$Pu, which have a half-life of about 24,000 years, one is talking about isolating these for nearly half a million years for twenty half-lives. That's an almost incomprehensible length of time for most of us, even for a geologist.

At present we have no demonstrated partitioning technology that would permit serious consideration of the options which would eliminate the waste

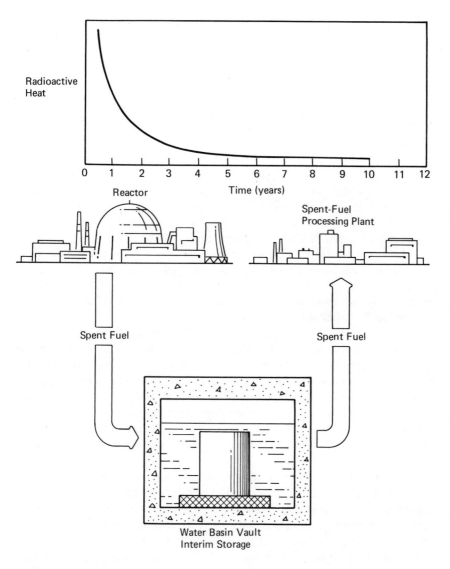

Source: ERDA 76-162, U.S. Energy Research and Development Administration.
**Figure 15-3.** Radioactive Heat Delay with Time.

from earth. It is doubtful that we will have such technology by the end of the century, and it may not be a viable option even after the year 2000.

Thus, the choices really narrow down to some means of isolation in geologic media. There are good reasons, which are not addressed in this presentation, that make seabed and ice sheet disposal unacceptable at this time; this discussion will

**Table 15-2**
**Alternate Options of Waste Isolation**

| |
|---|
| Elimination from earth |
|     Ejection by rocket |
|     Transmutation |
| Geologic isolation |
|     Seabed isolation |
|     Ice sheet isolation |
|     Deep continental isolation |

be confined to deep continental geologic isolation. A more detailed discussion and an evaluation of the several options can be found in the report of the American Physical Society study group on nuclear fuel cycles and waste management (*Review of Modern Physics,* January 1978).

The options for deep continental geologic isolation are indicated in table 15-3. Conventional mined cavities are those created by the same techniques as if one were mining an ore at depth. Here, however, we are concerned not with extracting something of value, but with disposing of something that we consider of no value. By definition, waste is of no value.

The possibility of developing solution-mined cavities for disposal of radioactive waste is viable only if the rock is soluble; consequently, rock salt would be the only medium that could be seriously considered for that. A matrix of drilled holes, whereby one would drill to depth and inject the waste in solid form into these holes, has the disadvantage of penetrating all geologic formations to the depth of the disposal horizon and thereby providing potential communication among these media from the surface down to the disposal level. Thus, it's not likely to be favored because sealing of these holes could present serious problems.

*Superdeep holes* refers to drilling large-diameter holes to depths of perhaps 20,000 feet or more and emplacing solid waste at the bottoms of these holes. This method could have great promise if one were to use a *single* hole to get to depth and then go off from that radially to produce horizontal holes into which one could emplace the waste canisters. Although the basic technology is already well established within the petroleum industry, additional work would be required to develop that technology for large-diameter holes to the desired depths.

A major consideration for any waste-isolation concept is a self-imposed requirement at present, known as *retrievability,* which states, in effect, that one must be able to retrieve the waste with essentially the same techniques and equipment that were used to emplace it. In point of fact, this criterion rules out all options except conventional mined cavities.

We can go back into a conventional mined cavity and retrieve the waste. We

**Table 15-3**
**Options for Deep Continental Geologic Isolation**

1. Conventional mined cavities
2. Solution-mined cavities
3. Matrix of drilled holes
4. Superdeep holes
5. Hydrofracture emplacement
6. Deep-well injection
7. Rock-melting concepts

could not do that very easily for the other options without significant expense and improved technology. For that reason the rest of the options are not discussed, although rock-melting concepts which have great potential for the future should be noted. Briefly, this simply means that one allows the concentration of wastes to be high enough that the heat produced by the radioactivity will locally melt the rock. When this subsequently cools, it essentially provides a barrier around that waste. If rock melt were coupled with superdeep holes, one could predict with great confidence that there would be no likelihood of return of that waste to the biosphere while it remained hazardous.

Present policy, however, supports storage and disposal in a conventional mined cavity. Emplacement of the waste would be in canisters like that shown in figure 15-4. The initially liquid high-level radioactive waste would be solidified and dispersed in a matrix within the canister. At present glass is the favored waste form (matrix), with about a 25 percent loading of waste to glass in a canister that's about 0.3 meter by 3 meters long. For a typical reactor capable of producing 1 gigawatt of electricity, 1000-megawatt capacity, this reactor would produce about ten canisters of high-level waste each year.

Figure 15-5 is an artist's conception of a conventional mined-cavity type of repository in which such canisters could be placed. They would be lowered through a shaft and put into storage rooms. One possibility would be to drill holes in the floor of the storage rooms on an even spacing and emplace the canisters in these holes. A principal concern here is that the distribution of heat produced by the waste be such that it would not, in some way, compromise the security of isolation.

While we are considering this concept of a repository, attention should be called to several aspects of the problem that are not always kept adequately separated.

There are really three periods of concern with regard to waste isolation. The first is the operational period of the repository which may be of the order of, say, 30, 50, or at most perhaps 100 years. That's a relatively short time, considering the time of desired isolation of the waste. The second period is the period during which the fission products (and attendant radioactive heat) are important, i.e., the first 1000 years. The third period is beyond that and extends perhaps beyond 1 million years.

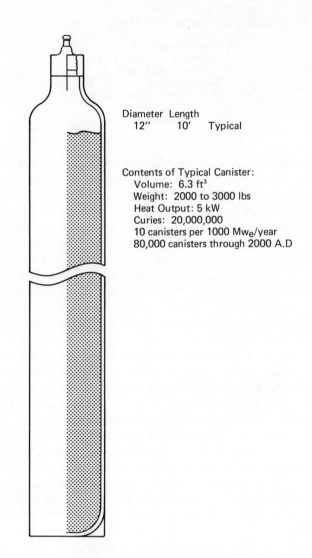

Diameter Length
   12"      10'      Typical


Contents of Typical Canister:
   Volume: 6.3 ft³
   Weight: 2000 to 3000 lbs
   Heat Output: 5 kW
   Curies: 20,000,000
   10 canisters per 1000 Mw$_e$/year
   80,000 canisters through 2000 A.D

**Figure 15-4.** "Typical" Canister.

A second aspect is the separation of effects in space as well as in time. Effects resulting from emplacement of the waste typically will be localized, e.g., the heat produced will have its greatest influence in the immediate vicinity of the repository. With time, the heat will affect a larger area, but the heat produced will also diminish with time. Thus, a spatial effect exists with local,

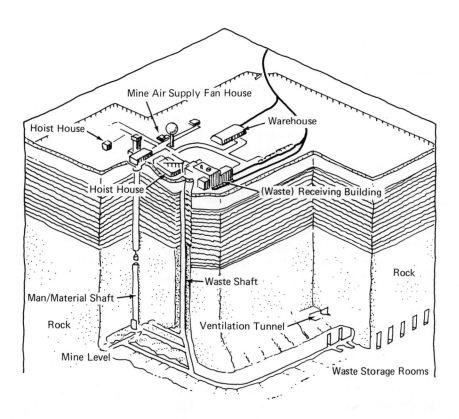

**Figure 15-5.** Simplified Cutaway of Repository.

near-field influence most likely. The possibility of far-field effects also exists, in some instances, particularly as it relates to fracturing and to the transport of groundwater, which will be discussed later.

It is important to recognize this spatial effect. What happens only in the vicinity of the repository may not affect what happens in the far field. More particularly, it may not influence transport of the waste away from the repository.

If we put waste into a repository of this or some similar design, where we have burial at a depth of 2000 to 3000 feet, how secure is this likely to be? The cavity will have to be back-filled, probably with the material that had been taken out, but it simply is not possible to fill it completely, and it can never be restored to essentially its initial state. There will always be some voids remaining. Moreover, there will be shafts and exploratory drill holes in the vicinity. All these will need to be sealed in some manner, but could potentially have some effect.

The question really is, Will the waste isolation be secure once the repository is sealed, or is there some possibility that the waste could move back to the biosphere?

There are only two ways in which the waste can get back to the biosphere. The first is to physically transport the rock mass in which the waste is emplaced back to the biosphere by some natural process or human activity. The second is to dissolve that waste in groundwater, which exists in all rocks to some extent, and move it with that groundwater back to the biosphere.

The next two tables outline possible ways in which the repository might be breached.

Table 15-4 lists natural breaching modes. Tectonic activity includes all types of movements in the earth and may simply result in uplifting the earth's surface, or it may lead to fracturing of the rocks and development of large discontinuities along which displacements occur. The latter are known as *faults*.

*Diapirism* refers to high mobility of a rock mass such that it flows and intrudes overlying rock units. That could be a possible problem with a rock such as salt. As is widely known, rock salt is one of the favored media being considered for a repository at this time.

Fracturing is important in that it may have an effect on the hydraulic transport properties of the rock. The presence of fractures may greatly increase the ability of the rock to transmit water. Fracturing could have far-reaching or far-field effects, as well as near-field effects in the vicinity of the repository.

Faulting could also affect the transport or flow of groundwater. Many people think the development of faults, or their presence, contributes to high

**Table 15-4**
**Natural Breaching Modes**

| |
|---|
| Tectonic activity |
| Fracturing |
| Faulting |
| Diapirism |
| Igneous activity |
| Erosion |
| Meteorite impact |

hydraulic conductivity, but faults can also provide a barrier to flow. Although most geologists would support the generalization that faulting in the area of a repository would be undesirable, one could argue that if the characteristics of a specific fault were well known, these could be utilized effectively in helping to protect the security of isolation of the wastes.

Almost no generalization can stand up to close scrutiny; in the matter of waste isolation, one must look at the important factors from all sides. Once a specific potential site has been identified, one can become very specific and evaluate the potential hazards associated with that particular site.

*Igneous activity* refers to all natural processes that involve molten rock. A concern here would be that molten rock at depth might possibly move upward toward the surface, intersect a repository, and carry radioactive waste back to man through ejection in a volcano or some other manner of eruption.

*Erosion* refers to the lowering of the earth's surface through the removal of soil and rock by natural processes operating at the surface. Exposure of waste in a repository by means of a direct meteorite impact is a highly remote possibility; the probability of this happening is so low as to be totally discounted.

Both erosion and meteorite impact can be discounted as serious factors if one were to put the repository deep enough; a couple of thousand feet would be adequate. By proper selection of the site with regard to depth and location, all the natural breaching modes can largely be discounted as significant factors.

The other category of breaching modes covers those that are man-induced (table 15-5). These include the possibility of sabotage, surface nuclear explosions, inadvertent penetration of the repository by drilling for resources such as water, and modifications related to mining excavation and emplacement of the waste.

Most of these possible breaching modes can be dealt with readily enough. For example, the repository could be placed at sufficient depth that surface nuclear explosions would not expose the waste.

As for sabotage, it should be obvious that once the repository is sealed, it would require a rather determined effort by any group to expose the waste. And, this would have to be done in the face of resistance by the controlling authorities.

Modifications to the repository environment related to waste emplacement lead, for the most part, to short-term and near-field effects. Stresses produced by

**Table 15-5**
**Anthropogenic Breaching Modes**

|  |
| --- |
| Sabotage |
| Surface nuclear explosions |
| Drilling |
| Modifications related to waste emplacement |

the excavation itself or by heat from the waste may lead to fracturing. This is basically an engineering problem, and it can be dealt with satisfactorily. To reduce the heat load, for example, one can simply decrease the amount of waste in individual canisters and increase the space between canisters in the repository.

Drilling is one activity that is impossible to predict. As a generalization, one can say that in the past, in searching for resources, man has always sought first those which are in concentration. Therefore, it would seem advisable to seek areas for repository sites in which there is no known significant concentration of some potentially useful resource.

Although we can't define with certainty what will constitute a resource tomorrow, we can say that it is likely that anything in concentration could be a potential resource. Therefore, if we choose a repository in an area where there is no such concentration, or where the material is sufficiently common that it would be sought elsewhere more likely than at depth, then the chances of inadvertent encounter of the repository by drilling would be significantly reduced.

Let us now return to the most critical question. If we put high-level radioactive waste in a repository of the conventional mined-cavity type, how can we be sure that the waste, once emplaced in that repository and once the repository is sealed, will indeed be secure?

We are talking about a period of at least 1000 years and, very likely, perhaps several hundred thousand years. It simply is not possible to predict with great certainty the natural and anthropogenic events that may occur within the next 100,000 years. However, we do know and understand well those physical and chemical processes most likely to be of concern with regard to this specific problem. What we can do is use computer models to analyze the situation and to evaluate specific factors that will influence the possible transport of the waste. Groundwater is the only medium potentially capable of moving the waste significant distances within the earth during the period of concern. If the waste should be dissolved, will it be transported back to the biosphere by groundwater?

Major efforts are currently under way to simulate the repository, characteristics of the geologic media, and relevant processes in order to determine what possible displacement of the waste may occur over long periods. In connection with the work of the American Physical Society study group on nuclear fuel cycles and waste management, we did some computer modeling to demonstrate a methodology that might be appropriate for the assessment of risk. More recently, the Nuclear Regulatory Commission in cooperation with Sandia Laboratories has been developing a methodology for risk assessment that utilizes even more sophisticated techniques of modeling to represent processes which could affect the waste repository.

Figure 15-6 is taken from the American Physical Society report and shows a representation of typical layered media in which a repository of the conven-

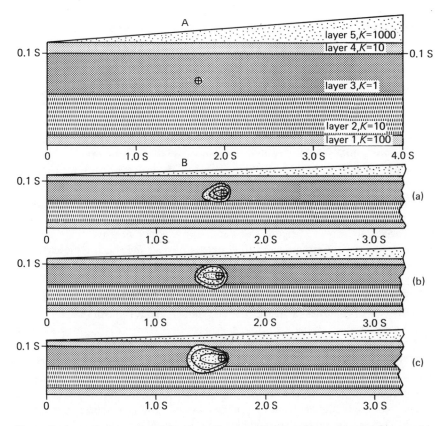

Note: Horizontally layered model with no-flow boundaries at sides and bottom; water-table gradient is $8.3 \times 10^{-3}$. Relative hydraulic conductivities of layers are indicated on figure; values assigned in the modeling results presented here are: layer $1 = 10^{-5}$; layer $2 = 10^{-6}$; layer $3 = 10^{-7}$; layer $4 = 10^{-6}$; and layer $5 = 10^{-4}$ cm/sec, respectively; and s = 11,420 m.

Note: The effect of increasing time on contaminant spread; (a) after 200,000 yrs; (b) after 400,000 yrs; (c) after 800,000 yrs. Hydraulic conductivities as given in A; dispersivity value assigned in all layers is 15 m; effective porosity is 10 percent. Contaminant source location indicated by ⊕. Intermediate contoured area includes between 2 and 10 percent of particles in system; the outermost contour encloses all particles in the system.

Source: "Report to the American Physical Society by the Study Group on Nuclear Fuel Cycles and Waste Management," *Rev. Mod. Phys.* 50:1, Part II (January 1978).

**Figure 15-6.** APS Study Group on Nuclear Fuel Cycles and Waste Management.

tional mined-cavity type might be placed. The three lower diagrams represent different periods of time—200,000 years, 400,000 years, and 800,000 years—after initial release of particles in the repository to the simulated groundwater system.

The model shows how far these particles would move with time for certain assumed characteristics of the natural system. If we knew precisely the characteristics of the system—the potential field that drives the groundwater, the boundary conditions, and the characteristics of the geologic media—then we could predict with considerable certainty where and how fast these particles will move.

Many uncertainties exist in dealing with a real system. There may be inadequate data on the characteristics and geology of the site, for example, to mention only the most obvious. We certainly are not able to predict precisely those events that will happen in the future, particularly as these relate to boundary conditions of the system. It cannot be stated with certainty whether we will have another period of glaciation, what changes might occur in rainfall and other climatic factors, or how much erosion will take place at the earth's surface, any of which might influence the hydraulic gradient. We can put some reasonable bounds on this, but we certainly cannot predict with certainty.

The results shown in figure 15-6 represent very conservative modeling. The driving potential for groundwater flow here is gravity, and this is represented by a typical water table gradient. It is very unlikely that natural gradients would ever exceed certain values, and a typical one is represented here.

The values of hydraulic parameters assigned to the model are also conservative—that is, properties assigned to the model are less favorable for waste disposal by at least one or two orders of magnitude than what might, with little difficulty, actually be found in natural situations.

The model is conservative further in that there is no restricted rate of release to the groundwater (leaching rate) and, once in the groundwater, particles are not sorbed on mineral species during the transport. Both factors could be important in reducing significantly possible concentrations in natural systems.

In this very conservative modeling, the particles are made available to the groundwater system and are transported at the rate at which the groundwater flows for the assumed media properties and boundary conditions. As stated, one could expect various dilution factors to be important in most geologic situations. In the diagrams of figure 15-6 there are three contours to show the percentage of particles distributed in the system. The intermediate contoured area contains between 2 and 10 percent of the particles; no particle lies outside the outermost contour. With increasing periods of time, the areal distribution increases; but one should note that even after 800,000 years, all the particles remain confined within a small zone, and have not even reached the higher units near the surface.

As for the presence of water, it is assumed that the rock layers are completely saturated, that movement of the groundwater is in response to a gravitational potential produced by the water table configuration, and that we have no-flow boundaries on both sides and at the base of the system. These are reasonable assumptions. It is possible, for example, for a geologist to define where no-flow boundaries are in a natural system, such that one could analyze a present situation. The problem comes in attempting to define some future situation. What might happen to the water table? What might happen to location of the boundaries? Nevertheless, one can place bounds or limits on these and get some feel for the possible consequences.

For this particular model the initial disposal was 600 meters. The exact depth and other dimensions of the model, for that matter, are not really very important because this is intended only to demonstrate a methodology. One could move the location of the waste repository up or down, but dimensions of the contaminant spread and its location within the system would depend on the properties that one assumes.

The model assumed a hydraulic conductivity, a principal parameter controlling the flow of groundwater, of $10^{-7}$ centimeters per second. This is one or two orders of magnitude faster than typical values for favored disposal media. In spite of the very conservative assumptions in this model, it is clear that limited transport has occurred. If more realistic values were used, there would presumably be even less transport.

The scientific and technological problems associated with the management of high-level radioactive wastes are soluble. Certain questions undoubtedly will be resolved on the basis of political, social, or perhaps economic considerations. Nevertheless, certain technical criteria need to be established that, if complied with, will ensure the security of isolation. And, these criteria must not be compromised by nontechnical considerations. It should be recognized, however, that no one site is likely to have to the fullest degree all the desirable characteristics one can identify. An optimal situation must be sought whereby if some characteristic associated with a potential repository site is not in itself desirable, there are other factors which, taken collectively, can make this an acceptable site.

We can dispose of high-level radioactive wastes safely. Whether we will dispose of these safely will depend on whether the issues are resolved on scientific or other bases.

# 16 The Fusion Hybrid

*Hans A. Bethe*

There is now great hope in the community that feasibility of fusion power will be established in the 1980s. If and when this comes about, the nation must decide how to use fusion power. One possibility is to use pure fusion which has the advantage, compared to fission, of considerably reduced radioactivity. However, it is likely that fusion reactors will be much more expensive than fission reactors. Factors between 1.5 and 5 in investment cost have been suggested by various authors, and maintenance costs are also likely to be higher. The nation will have to decide if it wants to pay that high a price for reduced radioactivity.

A second possibility is to use a hybrid, in which the fusion machine is surrounded by a blanket in which the fusion neutrons produce fission. A considerable increase, by a factor of 10 or more, in the power can be achieved in this manner. However, this type of reactor seems to combine many of the disadvantages of both fusion and fission, i.e., the greater complication of fusion and the radioactivity of fission.

There is, however, another type of hybrid. In this type, the fusion machine is surrounded by fertile, rather than fissile, material, so that fissile nuclides may be produced for future use in ordinary fission power reactors. This hybrid of the second type has always been attractive to me, and I have therefore looked at it more closely. When I began this closer look, I was entirely unaware of the extensive calculations and literature on the subject. A nice general discussion has been given by P. Fortescue[1] of General Atomic. Physics and design parameters are summarized by J.D. Lee,[2] and performance characteristics by D.J. Bender,[3] both of Livermore; B.W. Augenstein[4] of Rand has studied the economic aspects. Moses, and Abdel-Khalik[5] have pointed out the advantages of the fusion hybrid for preventing nuclear proliferation. An early summary was given by Maniscalco.[6] Further literature is given in the references quoted, but even this will not cover all the relevant papers. Industrial companies, including Westinghouse, General Electric and General Atomic, as well as EPRI, have done active work on the concept. When I talk of a fusion hybrid in this chapter, I mean one of this second kind, i.e., a fusion fuel factory (FFF).

For many reasons which will be discussed below, my preference for the fertile material in the fusion hybrid is thorium. It seems reasonable that every 14-megaelectronvolt neutron emerging from the fusion machine can produce about one atom of $^{233}U$ in thorium, plus one atom of tritium in the lithium which must also form part of the blanket. This estimate is based on the fact that a 14-megaelectronvolt neutron will generally first cause an $n,2n$ or even an $n,3n$ reaction in the heavy nucleus theorium, which makes two neutrons available for the two capture processes mentioned. Some 14-megaelectronvolt neutrons will cause fission in $^{232}$Th instead of the $n,2n$ reaction, and fission will produce about four neutrons.

Detailed calculations,[7] taking the diffusion of the neutrons in the blanket and more realistic geometries into account, give for the number of $^{233}$U formed in the blanket per tritium deuterium (TD) reaction

$$G = 0.5 \text{ to } 1 \tag{16.1}$$

$G$ stands for breeding gain. Since the fission of a $^{233}$U releases about 190 megaelectronvolt, while a TD reaction releases only 17.6 megaelectronvolt, the ratio of the "energy value" of the $^{233}$U produced, to the energy released in the TD reaction, is

$$r = 11G \tag{16.2}$$

It is this order-of-magnitude difference in the energy release in fusion and fission which will make the following calculations so favorable.

The most interesting advantage of the fusion hybrid is to help in preventing the proliferation of nuclear weapons. Like most people in the nuclear power community, I believe that the main task in preventing nuclear weapons proliferation must be diplomatic, political, and institutional (e.g., strengthening the IAEA, both in its power of inspection and in financial support). However, technical methods to minimize the contribution of nuclear reactors to the proliferation potential should be used if available.

A general scheme for this purpose has been outlined by H.A. Feiveson and T.B. Taylor.[8] They envisage a nuclear reactor economy composed of fast breeders and thermal neutron converters. In my proposal here, the fusion fuel factory would be substituted for the fast breeder. It would have a thorium blanket and thus would produce $^{233}$U. The FFF would be under international management and be located in politically stable countries, just like the fast breeders in the Taylor-Feiveson proposal. The converter reactors which use the $^{233}$U would be available to all countries and utilities which want them (dispersed reactors; of course the $^{233}$U will be denatured).

The great advantage of the FFF in this connection is that it can supply fuel for a large number of converter reactors. Assume the latter have a conversion ratio $C$. Then the ratio of the (thermal) energy produced in the converters to that produced in fusion is (see equation 16.1)

$$R = 11 \frac{G}{1 - C} \tag{16.3}$$

If the converters are light-water reactors, $C$ is about 0.6, so

$$R = 27G \tag{16.3a}$$

For advanced converters, like CANDU working with $^{233}U$ or the HTGR, $C$ can be as high as 0.9, so that

$$R = 110G \tag{16.3b}$$

In this latter case, then, one fusion hybrid with a fusion power of 1 gigawatt can supply (make up) fuel for 55 to 110 converters of 1 gigawatt each, using the estimate (16.1). For denatured fuel, the conversion ratio $C$ will be lower, perhaps only 0.8, so

$$R = 55G \tag{16.3c}$$

still a large number.

These ratios are independent of the price of the FFF. As will be shown in the next section, the economics of this device is not very favorable, but the advantage in the ratio $R$ will remain.

The large ratio $R$, from about 15 to about 100, is essential for the working of the Taylor-Feiveson scheme. The aim is to minimize the number of international, heavily guarded reactors and to maximize that of the "dispersed" reactors which can be located at places close to the consumers of electricity. The original Taylor-Feiveson scheme, using fast breeders (LMFBR) for the "international" reactors, is not very satisfactory in this respect: as Chang and collaborators[9] have shown, the ratio $R$ in this case is only about 3, even if advanced converters ($C = 0.9$) are used, and less than 1 in case of LWR.

The ratio $R$, as here defined, is not directly significant. Bender[10] introduces $R_t$, the ratio of the thermal energy produced by the converters, to that of the fusion hybrids. Roughly,

$$R_t = R/M' \tag{16.4}$$

where

$$M' = 1 + \frac{\text{fission power in the blanket}}{\text{fusion power}} \tag{16.5}$$

For a reasonable design with a thorium blanket, Bender gives $M' \approx 4$. Then $R_t$ will be only between 4 and 25.

The most important practical quantity, however, is the amount of electricity sold by the FFF to the grid. As will be shown below in the section on Performance, at least the early versions of the FFF will probably sell little electricity, if any. Thus the large $M'$ will mean only that a large amount of power will circulate in the FFF; the electricity sale will come from the dispersed reactors, which will be at suitable locations for this purpose.

Nevertheless, a small value of $M'$ seems desirable because dealing with large amounts of circulating heat is costly. Thorium is advantageous for this purpose because, even at 14-megaelectronvolt neutron energy, it has only a small fission cross section, about 0.36 barn. Since the total reaction cross section is likely to be about 3 barns, the probability of thorium fission is only 12 percent, the remainder being the $n,2n$ or $n,3n$ reaction. The thorium fission then contributes only relatively little energy, keeping $M'$ and therefore the cooling requirement low. The small fission cross section of thorium, usually a drawback, is an advantage in this case. In any case, as discussed above, the split between $n,2n$ reactions and fission will not appreciably affect the neutron balance. In a uranium blanket, with a larger fission cross section, $M'$ has been estimated[11] to be about 10.

The copious $n,2n$ reaction will produce $^{231}$Th which transforms into $^{231}$Pa which in its turn, by neutron capture and beta decay, becomes $^{232}$U. This material is well known for its strong gamma activity which will be a further safeguard against diversion for weapon manufacture; there will be more of this than in a thermal thorium-cycle reactor. Perhaps denaturing will then become unnecessary. Of course, the $^{232}$U will also make the use of the $^{233}$U for reactor fuel somewhat more troublesome, requiring remote fuel fabrication.


**Economics**

It is difficult to estimate the capital cost of a fusion hybrid because its design is so uncertain. However, figures of $2000 to $4000 per installed kilowatt of fusion power have been estimated.[12] Thus

$$K = \$3000 \pm 1000 \text{ per kilowatt (thermal)} \qquad (16.6)$$

Assuming 70 percent capacity factor, that is, 250 operating days per year, and 15 percent capital charge per year, this means a cost of

$$\$1800 \pm 600 \text{ per megawatt-day (thermal)} \qquad (16.7)$$

The number of grams of $^{233}$U produced per megawatt-day is given by the ratio $r$ in equation (16.2) since one megawatt-day of fission is equivalent to 1 gram of fissile material. Hence the cost of 1 gram of $^{233}$U from the FFF is

$$P = \$ \frac{165 \pm 55}{G} \qquad (16.8)$$

To this must be added the cost of chemical separation of uranium from thorium which we estimate at $10 per gram. (For the plutonium-uranium system, the cost per kilogram of heavy material is often given as $200, and we assume that chemical processing takes place when 2 percent of uranium has been built up in the thorium.) Assuming $G = 0.5$ to 1, the price per gram of $^{233}U$ is then

$$P' = \$120 \text{ to } \$450 \tag{16.8a}$$

This is very expensive. The present price of fuel is about $33 per gram of $^{235}U$ contained in 3 percent material; this is based on a cost of $40 per pound of $U_3O_8$ and $100 per unit of separative work. Now $^{233}U$ is somewhat more valuable than $^{235}U$, especially in an advanced converter, so the price [equation (16.8a)] may correspond to $100 to $375 per gram of $^{235}U$. Assuming the cost per separative work unit (SWU) to stay constant, this corresponds to about $180 to $700 per pound of $U_3O_8$. Especially the upper part of this range is not likely to be reached by raw uranium in the foreseeable future.

In fact, it is likely that Tennessee shale would be available at less than $200 per pound of $U_3O_8$, and even granite at less than $700. But the use of such resources would do considerable violence to the environment, and both re-sources are finite. The remaining alternative would be uranium from the ocean, an even more uncertain prospect than the FFF, especially with regard to cost.

On the other hand, there is the fast breeder. Dr. H. Hurwitz of the General Electric research laboratory has suggested the following price estimates: An advanced converter (CANDU or HTGR), with a conversion ratio $C = 0.9$, may cost about $200 more per kilowatt (electric) than an LWR with $C = 0.6$. Likewise, a fast breeder with $C = 1.2$ may cost about $200 more than an advanced converter. Using the same argument as in equations 16.6 to 16.8, we find, assuming 33 percent thermal efficiency, that

$$K_b = \$67 \text{ per kilowatt (thermal)} \tag{16.6a}$$
$$= \$40 \text{ per megawatt-day} \tag{16.7a}$$

Now 1 megawatt-day corresponds to the fission of 1 gram of fuel, and hence to 0.3-g difference in fuel consumption between advanced converters and LWR (or between breeders and advanced converters). Hence the effective cost per gram of fuel is

$$P = \$133 \tag{16.9}$$

To this we add $10 for reprocessing and get

$$P' \approx \$140 \tag{16.9a}$$

This price has an uncertainty proportional to that of the price difference of $200. Within these large uncertainties, the price of fuel from breeders is about the same as the lowest cost that can be expected from the fusion hybrids. But even the breeder-produced fuel would correspond to a price of about $250 per pound of $U_3O_8$.

In the comparison between FFF and LMFBR, we have to consider that (1) the LMFBR technology is rather well known while FFF is in an early stage of development, (2) LMFBR will probably produce nuclear fuel more cheaply, by a factor 1 to 3, than FFF, but (3) FFF lends itself to the establishment of a few, well-safeguarded, central fuel production centers which can supply a much larger number of dispersed reactors, while LMFBR does not. So we would have to pay the added price for an extra safeguard against proliferation.

Is the price acceptable? The net fuel consumed by the converter per megawatthour (electric) is about[a] $1/8$ $(1-C)$ grams where $C$ is again the conversion ratio. Hence the fuel cost per kilowatthour (electric) is

$$1.8 \ (1-C) \text{ cents for LMFBR}$$
$$(1.5 \text{ to } 5.5)(1-C) \text{ cents for FFF} \qquad (16.10)$$

Even with the highest price for FFF, and $C = 0.8$, this is only 1.1 cents, about the same as the present price of coal; for the "average" price of 3.5 and an average $C = 0.85$, it is only 0.5 cents. This compares with perhaps 2.5 cents per kWhe from the initial investment in an advanced converter (whose cost we assume to be $1000 per kilowatt (electric), including interest charges during construction, but excluding inflation escalation). These numbers are consistent with those of Bender who finds, with similar assumptions, that the fuel from FFF may add 25 percent to the cost of the fission reactor. So it seems to me that either the LMFBR or the FFF is acceptable in a mixed economy if advanced converters are used. As Augenstein has pointed out, the result is fairly insensitive to the price of the FFF, which is of course due to the small number of FFFs in a mixed fusion-fission economy.

## Performance of the Fusion Reactor

Bender has discussed in detail the electric energy produced and consumed by a fusion hybrid. The most important "figure of merit" of a fusion plant is the ratio

$$Q_p = \frac{\text{fusion power}}{\text{power input to heat the plasma}} \qquad (16.11)$$

---

[a]Using a thermal efficiency of 1/3 and 24 hours per day.

The fusion hybrid will be self-sustaining electrically if

$$(Q_p M' + 1)\, \eta_h \eta_{tb} = 1 \qquad\qquad (16.12)$$

where $\eta_h$ is the efficiency of heating the plasma by the input electric power, $\eta_{tb}$ the efficiency of conversion of heat from the fusion reactor into electricity, and $M'$ is given by 16.5. The extra 1 on the left of 16.12 is the heat in the plasma.

Typical values, given by Bender, are

$$\eta_h = 0.5 \qquad \eta_{tb} = 0.35 \qquad M' = 4 \qquad\qquad (16.12a)$$

Then the $Q_p$ required to satisfy 16.12 is

$$Q_p^* = 1.2 \qquad\qquad (16.13)$$

This is a rather modest performance of the fusion reactor, close to that projected for the Princeton Tokomak, TFTR, which is scheduled to begin operation in 1981. The Livermore mirror machine, MFTF, scheduled also for 1981, is projected to achieve only $Q_p = 0.1$, but its extension to the tandem design, with a long cylindrical plasma cylinder between two mirrors, should do much better and would have a particularly suitable geometry for a breeding blanket. Electrical self-sufficiency is, of course, not necessary, but it is useful. On the other hand, it may be cheaper to build a fusion hybrid in which the entire heat is rejected, and this might more than compensate for the cost of the electricity which such a device would have to obtain from the outside.

Condition 16.13 is close to "scientific breakeven," $Q_p = 1$. Such a value of $Q_p$ was also assumed in calculating the cost per kilowatt of fusion power, 16.6. For other values of $Q_p$, the cost 16.6 should be multiplied by

$$K' = \frac{M' + 1/Q_p}{M' + 1} \qquad\qquad (16.6b)$$

Thus a high fusion performance $Q_p$ does not make much difference, but a very low $Q_p$, like 0.1, will increase the cost greatly.

Once a satisfactory $Q_p$ has been achieved, the most important factor in the economics is the cost of the heat removal from the fusion hybrid. This can probably be reduced most effectively by actually building fusion hybrids and improving them on the basis of experience. Ultimately, the price of fuel produced from them might compete favorably with that from LMFBR.

The hybrid could help greatly toward the orderly introduction of fusion into the economy. Even if it is presently not competitive in cost, the hybrid is closer to that goal than a pure-fusion device. Most importantly, only a relatively small number of hybrids need to be built to fuel the fission reactors. Assuming the United States needs 600 gigawatts (electric) of fission reactors by 2020, and assuming equation 16.3$b$ with $G = 0.65$, the makeup fuel for the fission reactors can be supplied by only 10 gigawatts (electric) [or 30 gigawatts (thermal)] of fusion power. The deployment of these fusion hybrids need not take very long. If the feasibility of fusion is proved in the mid-1980s, perhaps the first industrial hybrid could work before 2020, and the (perhaps 30) fusion hybrids required for fuel supply might be in place by 2040.

As experience is gained with fusion hybrids, the fusion reactor itself may improve; in particular, $Q_p$ is likely to increase, let us say to 10 or more. And at that point pure-fusion devices may become feasible and ultimately economical. The Soviets have for some time chosen this approach to fusion.


**Discussion**

The possibility of a fusion hybrid does not make the fast breeder unnecessary. The technology for a fast breeder is well known; that for the fusion hybrid is still in its infancy. The cost estimates for the fusion hybrid are very uncertain compared with those for the LMFBR. In the beginning, almost certainly, fuel from the fusion hybrid will be more expensive than from the fast breeder. In the course of time, this may change. Because of the great uncertainty in the estimate of the uranium resources, in both the United States and the world, I believe it is necessary to fully develop the LMFBR so that it is available for commercial use if and when needed.

Whatever method is used to stretch the uranium fuel supply (advanced converters, fast breeders, fusion hybrids, still other devices, or a combination of several of these), it seems to me that reprocessing of the fuel will be necessary. To obtain good performance from an advanced converter, it is essential to eliminate, by reprocessing, most of the neutron-absorbing fission products. While it is possible to irradiate the fuel elements for a converter reactor in the blanket of a fusion hybrid,[13] the neutron-conserving performance of both will surely be improved[14] by reprocessing and refabrication. In addition, the CIVEX process, developed by EPRI and the Harwell Laboratory, never completely separates plutonium from uranium and fission products, so that no possible bomb material ever becomes available in the chemical factory. A similar process can probably be developed for the thorium-uranium cycle. President Carter's emphasis on avoiding weapons proliferation in the nuclear fuel cycle has stimulated much useful work which will help to attain this aim, but once a satisfactory method for chemical separation has been developed, it should be made part of the fuel cycle.

I am very grateful to many people for discussions, for making their work available to me, and for correspondence. Due to this help, I was able to correct many errors.[15] My special thanks go to D.J. Bender and J.D. Lee of Livermore and their colleagues, to Henry Hurwitz of General Electric (Schenectady), Peter Fortescue of General Atomic, S. Locke Bogart of DoE, and E.L. Zebroski of EPRI. Important points were also brought up by W.G. Davey of Los Alamos, T.J. Connolly of Stanford University, R.N. Kostoff of DoE, and a number of others. My thanks to them all.

## Notes

1. P. Fortescue, *Nucl. Eng. Int.* 21 (1976):71; *Ann. Nucl. Energy* 2 (1975):29; *Science* 196 (1977):1326.

2. J.D. Lee, Second DMFE Hybrid Reactor Meeting, Washington, D.C., November 1977, preprint UCRL-80651.

3. D.J. Bender, Third ANS Topical Meeting on Controlled Thermonuclear Fusion, Santa Fe, NM, May 1978, preprint UCRL-80589.

4. B.W. Augenstein, Rand Corp. Report P-6047, December 1977.

5. R.W. Conn, G.A. Moses, and S.I. Abdel-Khalik, University of Wisconsin preprint.

6. J. Maniscalco, *Nuclear Technology* 28 (1976):98.

7. Lee, preprint UCRL-80651.

8. H.A. Feiveson and T.B. Taylor, *Bull. Atomic Scientists* 32, No. 10 (December 1976):14.

9. Y.I. Chang, C.E. Till, R.R. Rudolph, J.R. Deen, and M.J. King, "Alternative Fuel Cycle Options: Performance Characteristics and Impact on Nuclear Power Growth Potential," RSS-TM-4, Argonne National Laboratory, July 1977.

10. Bender, preprint UCRL-80589.

11. Lee, preprint UCRL-80651; Bender, preprint UCRL-80589.

12. Bender estimates the cost per unit of *total* thermal power of the fusion hybrid to be $2\pm1$ times that of a fission reactor, which is about $300 per kilowatt (thermal). But the total thermal power in his design is about 5 times the fusion power, so that the mean estimate per unit fusion power is $2 \cdot 5 \cdot \$300$ per kilowatt (thermal).

13. Bender, Lee, and Moir, Livermore report UCID-17607; Conn, Moses, and Abdel-Khalik, University of Wisconsin preprint.

14. Conn, Moses, and Abdel-Khalik, ibid.

15. A preliminary version of this chapter was circulated to many people in the field. In this, I was too optimistic, by about an order of magnitude, about the cost of fuel from a fusion hybrid. I apologize for the widespread confusion I caused by this preliminary version. (The original version of this chapter appeared in *Nuclear News,* May 1978.)

# Part V
# Conclusion

# 17 Bankrupt Energy Policy: The Abdication of American Leadership

## Robert R. Nathan

The subject matter chosen today reflects my feeling with respect to what I call the bankruptcy of America's energy policy. As an economist, and especially as one concerned with international matters, and as one concerned with economic strength of this country, I believe we are faced with a very serious energy situation.

Until a relatively few years ago, the United States was one of the real major powers. Without going back into the history of the economic development of this country and the role the United States played in the development of modern technology and modern industrialization, I would emphasize that it wasn't until after the turn of the century that this country became a creditor rather than a debtor nation, where we had substantial exports beyond those of any other country. Actually, it was World War I that gave a tremendous shot in the arm to America's industrial development and resource exploitation and put us ahead of most countries. Then came World War II, when so many industrial nations were battered, beaten, and destroyed. We built our strength to the point where at the end of World War II, we had more aluminum capacity, more steel capacity, more machine tool capacity, more industrial capability than we had when we went into the war. The result of this was our emergence as a major power.

In early November 1977, I attended a meeting of Asian countries, the United States, and Canada in Australia. At that conference, one of the three subjects had to do with America's energy policy and the implication of America's energy policy for international relations, international status, international responsibilities, and the international performance of the United States.

I walked away from the conference in Canberra, Australia, very, very much chagrined at the degree of bitterness and disrespect for the United States, the country which had played such a tremendous role in the early years of post-World War II in helping the destroyed industrial nations of the world to become rehabilitated and in helping, for the first time in all human history, the less developed countries of the world truly to begin the process of modernization and raising productivity and living standards.

The feeling at the conference was that the United States is betraying many of the less developed countries and the other countries in the world that are dependent on energy almost entirely from external sources. The expressions

143

generally went, "Well, you rely on foreign sources for only half or nearly half of your oil requirements, where we depend 100 percent on imported energy sources." And "The United States still had vast production capabilities in oil." And "In terms of natural gas, the United States is still largely independent." This was from Korea, Japan, the Philippines, and several other countries as well. And what they said is true. We rely today almost entirely on our own resources for natural gas, although we are now placing commitments abroad for liquified natural gas (LNG) in Algeria and Indonesia, where gas is converted to liquid form, transported here, and then reconverted back to gaseous form in the United States. But that's really such an infinitesimal segment of our total gas consumption that one can say at present, subject, of course, to limitations in the use of natural gas, that we still are largely self-sufficient.

These countries believe that the United States, as a wasteful user of energy—with 6 percent of the world's population using a multiple of many times that ratio of total energy—is neglectful of the well-being, security, and continuity of economic progress of the rest of the world.

This perception wasn't entirely new to me because I had been talking, studying, and analyzing American economic and energy problems. And I couldn't very well say, "Well, be patient with us. It takes us a little time. A democracy doesn't always work very rapidly."

They couldn't understand the differences between Carter and the Congress. They couldn't understand what was really going on in terms of the conflict between different energy groups in the United States. They couldn't understand why we weren't conserving. And they couldn't understand why we weren't spending a lot of money to develop new sources of energy, on research and development, and on trying to do in the energy field what we had accomplished when compared to reaching the moon and developing atomic weapons in World War II. They evidenced a considerable degree of frustration and disrespect, really, and anger toward the United States.

Let's take a serious, quick look at this picture, and let me try to explain why our performance has been so dismal.

First, it's now over four years since the oil embargo in the fall of 1973. It wasn't that we didn't have warning prior to 1973 that the United States was facing a rapid depletion of resources or at least using up a limited resource. I recall—I think it was before 1973—a statement from one of the great geologists in this country that it looked like the fossil fuel built up over 4 billion years was going to last only about 200 years. That's a rather sobering thought. And there were warnings in the 1950s and 1960s.

Here we are, more than four years after the oil embargo altered us to the realization of the possibilities and the challenges, and I must frankly say that we have done precious little in terms of meeting this problem. I don't mean that nothing has been done; but, by and large, considering the nature and magnitude of the problem, not a great deal has been done.

Now, let me not be too general on the nature of the problem. I'm not going to state that we are faced with a crisis that is three or five or seven years hence. Having read the report by Carl Wilson with the M.I.T. group and some of the technicians in other countries, the report of the CIA and others about whether we're going to have an oil shortage in 1982 or 1985 or 1990, I know there are different views.

The issue is not whether we're going to have a critical problem in three years or five years or ten. The real issue is that we are in the process, because of the way we're going, of running out of oil. And it will probably be a gradual process—as supplies begin to get short, the flow isn't there, the pressure isn't there, secondary recovery doesn't occur, and the new technologies don't evolve rapidly. It's not going to be 22 1/2 years and then a sudden drop-off. We don't know when the toughest pinches are going to come.

Dr. Arnold Safer of Irving Trust has written a very interesting and optimistic paper projecting individual country prospects for increased oil production. There are good possibilities in Mexico, the North Sea, the north slope of Alaska, and other locations including offshore oil in the United States. But the basic picture is that we are using up reserves very rapidly. And while discoveries from 1973 to 1976 increased our reserves by about 5 billion barrels a day, as Dr. Safer pointed out, it's only about a 2 percent increase in reserves a year, in response to a 400 or 500 percent rise in prices.

If we relax and say the marketplace is going to solve this problem, we're kidding ourselves into a very much more serious set of problems than we have now. It's amazing how many people will say that if we get rid of all the controls, the marketplace will solve it; you'll have elasticity of responsive supply with higher prices, and you'll have elasticity of slowing down of demand with higher prices.

We ought to use the marketplace in our economy and give it the greatest force we can, but not all problems can be solved in a marketplace which doesn't function very well under many circumstances.

We've had a high degree of inflation the last four years, with unemployment and excess industrial capacity. Look at some of the industries. The price of automobiles rose just as fast when there were 7 million cars produced a year as when there were 11 million. In other words, you just don't have that nice Milton Friedman model of responsiveness of the marketplace to devise a supply-and-demand situation that solves all the problems.

And, therefore, while I'd like to see us resort to the fullest on the mechanism that the marketplace has afforded, we shouldn't tell ourselves this problem is going to be solved by just relying on the marketplace.

Americans for Energy Independence (AFEI) has pursued the solution, or at least attacked the problem, through three general approaches. One is conservation; but we have no illusions that conservation is going to solve our problem. Conservation is difficult, especially in a society where people are horribly spoiled and hate any kind of inconvenience. We love to talk about voluntary responses—which don't happen to be effective enough—we hate regulation. We hate standards. We hate trying to set conditions in many, many ways which would bring about conservation more rapidly. But, nonetheless, Americans for Energy Independence feels very strongly that conservation can give us breathing time. It is also terribly important for international purposes to give a sense that we are jointly responsible and responsive, along with other countries which are very worried about the situation.

Conservation is also a very important mechanism to get people involved, to get people to understand. Too many Americans don't believe that there is an energy problem. If we can begin to get a better understanding of the importance of conservation and citizens participate in conservation efforts, it will have a wholesome effect in terms of conception and grasp of the nature of the problem.

Second, we believe it is terribly important to use the resources we have. Develop and use our oil and gas resources to the fullest extent and rapidly expand coal utilization. Move forward with nuclear energy. We can clean up whatever problems there may be in the use of nuclear power. And then, above all, make use of research and development of the intensive nature that we have used in the past when we were faced with problems.

Now, let us examine briefly the domestic and international problems that make me so concerned.

First, on the domestic front, we have had twelve years of serious inflation. Inflation started in this country in 1965-1966 and now it's 1978. For twelve years, we've had the worst peacetime inflation the United States has known. And this peacetime inflation has been aggravated in a very considerable degree by the energy situation.

As far as I can see, no matter how restrained OPEC is, and even if we do what Dr. Safer proposes (try to negotiate with the OPEC countries, taking advantage of the temporary surplus of supplies relative to immediate demand), I still think that the longer pull on the energy side is going to be inflationary.

And if you agree with me that inflation is a kind of cancer on our body economic, then you realize and agree that something effective has to be done about energy to try to restrain the demand side without massive unemployment, zero growth, and things of that nature, which I'm convinced we can't live with, and, on the other side, to try to expand our supplies.

Unfortunately—and this is not unrelated to energy because it's all part and parcel of a set of bankrupt policies—we've been trying to fight inflation with unemployment. That's been our major vehicle for trying to bring inflation under control. It hasn't worked, and I don't think it's going to work any better in the future. We've tried it now for several years. The wholesale price index rose 0.7 percent in December; it rose 0.4 in November and 0.8 in October. In these three months, the last quarter of 1977, the annual rate of increase was almost 8 percent. That's not price stability. We had a good break in December in unemployment, but we should not be surprised if it goes up in January because the statistical quirks of an index for one month are something that you can't be too sanguine about.

The energy situation has given us a degree of mental paralysis, a lack of vigor and dynamism and positiveness in economic policies. And one can't exaggerate the importance of this situation. We have very serious problems here in deciding what to do with the economy to meet this situation in energy, which played a part in bringing about the double-digit inflation of 1974 and 1975 and which continues to be a force.

The whole economy is slack because of the many uncertainties. One of the grave difficulties that the United States faces today is the lack of a high level of investment. We aren't modernizing. Look at the steel industry or whatever industry you want. You can look at area after area and we're not investing in new plant and equipment. We're not updating. We're not expanding capacity. We're not making the economy more vigorously competitive. And in a very considerable degree, this lack of vigorous investment policies' measures and responses stems from the uncertainty about the whole energy picture.

It is distressing to read what some economists say—that in the next ten years we're going to need so many hundreds of billions of dollars of investment in the energy field and we're going to have an inadequate source of savings to finance this expansion.

The record indicates just the opposite. We seem to have too much savings at high levels of activity, and we don't have the private demand to use up those savings; so we have government deficits. The offsets to savings in the private side are not dynamic and vigorous enough. This is why we have those huge deficits. Because we don't have a vigorous economy, we have recessionary deficits which arise out of a lack of revenues because there is idle capacity in all levels of GNP.

And the result is that we are not investing in modernization. And whether it is steel or other industries that are having great problems with competition, or whether the problem is facing up to and meeting energy problems, we seem to have a kind of paralysis which, I regret to say, stems in some measure from our energy uncertainty. You have a vicious circle: Energy uncertainties stimulate fear and uncertainties on the part of people who hold back on investments. Industries say, "We've got to protect ourselves" and opt for protectionism, which is self-defeating in my judgment.

Our whole economic policy and stagflation in some measure is attributable, I believe, to inadequacy, or lack of successful performance on the energy front.

Now, let me turn briefly to the international aspects of the energy problem. Last year, the United States spent approximately $45 billion to import oil. Almost half—47 to 48 percent—of our oil supply came from abroad. Even with all the great productive capacity of this country, all the genius of people in our universities, research laboratories, and industry, the result was that we had a balance of trade deficit last year of almost $30 billion.

That means we imported almost $30 billion more than we exported. That's one of the major reasons why the dollar is weak. And the weakness of the dollar, in turn, has implications for interest rates and investment prospects. You end up with a policy to fight inflation with unemployment. One can understand why common stocks are not any longer a good hedge against inflation, which was traditionally true in this country.

Until now, we've been able to manage financially by capital flow. What I mean by capital flow is that a lot of the OPEC countries find that the United States is not the best place in the world, but it isn't a bad place to invest. So they deposit money in our good banks, and they buy government securities or even A securities, land, businesses, and stock in banks. But what happens to the less developed countries (LDCs) of the world? What happens to the countries that don't have domestic resources of energy and need to import oil but don't have the means?

I'm proud of the continuation of a United States effort to try to help the less developed countries. Although our flow of capital to the LDCs is less than it was, we are still trying to recycle some of this flow from the OPEC countries into the United States to get it back into the hands of the less developed countries. We take in the funds from the OPEC countries short term, 90 days, or on notice of withdrawal, assessing it out, or all AAA or AA or A securities. Then we turn around and lend our BBs or CCCs. The banks have done this, too, and I think some of them are a little bit concerned. We may have overextended some of our loans. You don't have to be a banker to realize that if you take in billions of dollars that can be withdrawn overnight, and you lend it out for five, ten, twenty, or thirty years, that creates a very precarious situation. I'm strongly in favor of helping the less developed countries, but we've got to do this mainly through the monetary fund, the World Bank, the regional banks, and inter-governmental transactions, rather than through private channels.

The net result is that our economy on the whole has become much more jittery, uncertain, insecure. Unless we come up with an energy program and policy that embodies the three principles which Americans for Energy Independence really pushes—namely, a really meaningful conservation program, a very substantial increase in energy production for resources that are currently available, and then a major R&D program for the years ahead—we are going to be in trouble. Again, I'm not going to argue whether it's five, eight, fifteen, or twenty years. We have a limited time, and it will take time to develop these solutions.

In other words, unless we come up with an effective energy policy, the prestige, status, and influence of the United States will decrease, which could be quite tragic for the free countries of the world.

Let us consider the growing indebtedness of the less developed countries that are not oil exporters.

Not many people really realize what's happened in the whole development process around the world. Ten years ago, Korea was the showcase of economic development, and in many ways it is today. Korea has some coal, but not much. As a matter of fact, the coal there is very soft and brittle, and they have to make it into briquettes. They have no oil, and they have no natural gas. The result is that they're beginning to face very grave problems. They have been able to handle the situation because their labor on the whole is quite disciplined. They're able to pay rather modest wages. And they've developed very high productivity. Their export trade is pursued vigorously and very successfully.

However, I am convinced that if this oil and energy situation isn't corrected to some degree, we're going to face in this country the worst protectionism we've had in over forty years. You see it now. You see it in the pressure to cut down on imports of this and that. If that happens, the developing countries of the world are in very real trouble. And don't think they aren't aware of it.

And while you might say Korea ought to be very respectful and appreciative of what we've done to help them in the last twenty years, the big issue is, What did you do for me yesterday? and What are you going to do for me tomorrow? People are desperately trying to move ahead. I see a conflict here involving those countries which are big oil exporters. Many of them are spending as much as they possibly can and still building up big reserves.

One might say, Why don't Saudi Arabia, Kuwait, Abu Dhabi, and others pour out the billions to the less developed countries? The answer is that they don't have the know-how that goes with capital flow. Don't forget that under our aid program, our Point Four program, our P.L.-480 program, and the funding that went through the international institutions all were associated with capital resource or finance know-how and technical abilities. The OPEC countries don't have that. Whole new relationships are evolving, and they're uncertain about their role in the world and their own future.

I believe that if the United States were to look inward and say, "We're not going to take this money that's coming in to us in a precarious, short-term kind of way with low risk and put it out at high-risk loan terms," we could have economic chaos and the slowdown of development in many critical parts of the world.

We need, I believe, some new kinds of organizations, international instruments, and United States development policy. But until we come up with sound energy programs and policies, I don't see that we're going to be able to come up with development policies, stabilization policies, and the like.

In conclusion, I'm not going to try to analyze every piece of President Carter's message. His general statement on energy last spring was quite good. It was tough. It was harsh. But, frankly, the legislation was not as tough as the rhetoric. And even with the legislation being weaker than the rhetoric, we're not making much progress.

In the United States Congress today, we are facing a set of specific interests. The oil states want this, and the gas states want that, and these states want this.

We're not facing up to the issues that are confronting this country and the world. Until now, the conflicts have arisen primarily because there hasn't been a real national determination or national commitment to meet what I regard as an emerging crisis.

I hope that through meetings and conferences like this, the people in the Congress will come to understand. I think, they need it even more than the administration, although some education on the administration side wouldn't hurt either. I believe that through expressions of public concern, public uprisings, through positive, constructive support, demonstrating a realization of what the challenges are and the severity of the issue, we can pull ourselves out of this morass. So far the progress that has been made is very small relative to the magnitude of the challenge.

# About the Contributors

**Hans A. Bethe** is a professor emeritus of physics at Cornell University and a Nobel Laureate.

**Robert Williams** is senior research scientist and member of the program for nuclear policy alternatives at the Princeton University Center for Environmental Studies.

**Macauley Whiting** is a consultant for The Dow Chemical Corporation.

**Shirley Sutton** is executive director of Project Pacesetter in Pittsburgh.

**Clarke Watson** is board chairman of the Westland Energy Resource Development Corporation.

**Arnold Moore** is director of the Federal Agencies Department at the American Petroleum Institute.

**K.S. Feindler** is technical director at Grumman Ecosystems Corporation in Melville, New York.

**Arthur M. Squires** is a Frank C. Vilbrandt Professor of Chemical Engineering at Virginia Polytechnic Institute and State University and is a well-known specialist in coal.

**Carl Bagge** is president of the National Coal Association in Washington, D.C.

**Richard Disbrow** is vice chairman, administration, of the American Electric Power Service Corporation.

**Frederick N. Ferguson** is deputy solicitor at the Department of the Interior.

**Aubrey Wagner** is chairman of the board of the Tennessee Valley Authority.

**Charles Till** is the director of the applied physics division of the Argonne National Laboratory.

**John Simpson,** until his retirement in 1977, was chairman of the energy committee of the board of directors of the Westinghouse Corporation.

**Fred Donath** is a professor of geology at the University of Illinois at Urbana-Champaign.

**Robert R. Nathan** is a consulting economist in Washington, D.C.

# About the Authors

**Elihu Bergman** is the executive director of Americans for Energy Independence, a nonprofit public interest coalition based in Washington. He is an international development specialist, whose professional experience has included assignments in the United States and abroad with government, private, and academic institutions and foundations.

Prior to assuming his current responsibilities, Dr. Bergman served as assistant director of the Harvard Center for Population Studies. He has published extensively on the political analysis of population policy making.

Dr. Bergman did undergraduate work at Reed College and received graduate degrees at the University of Chicago and the University of North Carolina at Chapel Hill.

**Hans A. Bethe** was awarded the Nobel Prize in physics in 1967 for his work in explaining the mechanism by which we obtain energy from the sun. Professor Bethe taught at Cornell University from 1935 to 1975, and continues his association with the Laboratory of Nuclear Studies at Cornell.

During World War II, Professor Bethe served as leader of the theoretical physics division at Los Alamos Laboratory.

Professor Bethe is chairman of the board of Americans for Energy Independence. He has served in consulting capacities with government and private institutions and is the recipient of numerous awards for his scientific achievements, including the Presidential Medal of Merit in 1946 and the Enrico Fermi Award in 1961.

Professor Bethe was born and educated in Germany and came to the United States in 1933.

**Robert E. Marshak** is president of the City College of New York. A member of the University of Rochester faculty from 1939 to 1970, Dr. Marshak served as chairman of the Department of Physics and Astronomy, and as a Distinguished University Professor. He has taught and conducted research at scientific institutions in the United States and abroad.

Dr. Marshak has been active in numerous scientific societies at the national and international levels, including the National Academy of Science, the American Academy of Arts and Science, and the National Commission for UNESCO. He is now a member of the advisory council of Americans for Energy Independence.

Dr. Marshak did his undergraduate work at Columbia College and graduate work at Cornell University, where he earned the Ph.D. under the direction of Hans A. Bethe.